Lecture Notes in Control and Information Sciences

Edited by A. V. Balakrishnan and M. Thoma

Vol. 1: Distributed Parameter Systems: Modelling and Identification
Proceedings of the IFIP Working Conference, Rome, Italy, June 21–26, 1976
Edited by A. Ruberti
V, 458 pages. 1978

Vol. 2: New Trends in Systems Analysis
International Symposium, Versailles, December 13–17, 1976
Edited by A. Bensoussan and J. L. Lions
VII, 759 pages. 1977

Vol. 3: Differential Games and Applications
Proceedings of a Workshop, Enschede, Netherlands, March 16–25, 1977
Edited by P. Hagedorn, H. W. Knobloch, and G. J. Olsder
XII, 236 pages. 1977

Vol. 4: M. A. Crane, A. J. Lemoine
An Introduction to the Regenerative Method for Simulation Analysis
VII, 111 pages. 1977

Vol. 5: David J. Clements, Brian D. O. Anderson
Singular Optimal Control: The Linear Quadratic Problem
V, 93 pages. 1978

Vol. 6: Optimization Techniques
Proceedings of the 8th IFIP Conference on Optimization Techniques, Würzburg, September 5–9, 1977
Part 1
Edited by J. Stoer
XIII, 528 pages. 1978

Vol. 7: Optimization Techniques
Proceedings of the 8th IFIP Conference on Optimization Techniques, Würzburg, September 5–9, 1977
Part 2
Edited by J. Stoer
XIII, 512 pages. 1978

Vol. 8: R. F. Curtain, A. J. Pritchard
Infinite Dimensional Linear Systems Theory
VII, 298 pages. 1978

Vol. 9: Y. M. El-Fattah, C. Foulard
Learning Systems:
Decision, Simulation, and Control
VII, 119 pages. 1978

Vol. 10: J. M. Maciejowski
The Modelling of Systems with Small Observation Sets
VII, 241 pages. 1978

Vol. 11: Y. Sewaragi, T. Soeda, S. Omatu
Modelling, Estimation, and Their Applications for Distributed Parameter Systems
VI, 269 pages. 1978

Vol. 12: I. Postlethwaite, A. G. J. McFarlane
A Complex Variable Approach to the Analysis of Linear Multivariable Feedback Systems
IV, 177 pages. 1979

Vol. 13: E. D. Sontag
Polynomial Response Maps
VIII, 168 pages. 1979

Vol. 14: International Symposium on Systems Optimization and Analysis
Rocquentcourt, December 11–13, 1978; IRIA LABORIA
Edited by A. Bensoussan and J. Lions
VIII, 332 pages. 1979

Vol. 15: Semi-Infinite Programming
Proceedings of a Workshop, Bad Honnef, August 30 – September 1, 1978
V, 180 pages. 1979

Vol. 16: Stochastic Control Theory and Stochastic Differential Systems
Proceedings of a Workshop of the „Sonderforschungsbereich 72 der Deutschen Forschungsgemeinschaft an der Universität Bonn" which took place in January 1979 at Bad Honnef
VIII, 615 pages. 1979

Vol. 17: O. I. Franksen, P. Falster, F. J. Evans
Qualitative Aspects of Large Scale Systems
Developing Design Rules Using APL
XII, 119 pages. 1979

Vol. 18: Modelling and Optimization of Complex Systems
Proceedings of the IFIP-TC 7 Working Conference Novosibirsk, USSR, 3–9 July, 1978
Edited by G. I. Marchuk
VI, 293 pages. 1979

Vol. 19: Global and Large Scale System Models
Proceedings of the Center for Advanced Studies (CAS) International Summer Seminar
Dubrovnik, Yugoslavia, August 21–26, 1978
Edited by B. Lazarević
VIII, 232 pages. 1979

Vol. 20: B. Egardt
Stability of Adaptive Controllers
V, 158 pages. 1979

Vol. 21: Martin B. Zarrop
Optimal Experiment Design for Dynamic System Identification
X, 197 pages. 1979

For further listing of published volumes please turn over to inside of back cover.

Lecture Notes in Control and Information Sciences

Edited by A.V. Balakrishnan and M. Thoma

29

M. Vidyasagar

Input-Output Analysis of Large-Scale Interconnected Systems

Decomposition, Well-Posedness and Stability

Springer-Verlag Berlin Heidelberg GmbH 1981

Author
Prof. M. Vidyasagar
Dept. of Electrical Engineering
University of Waterloo
Waterloo, Ontario
Canada

ISBN 978-3-540-10501-5 ISBN 978-3-540-38505-9 (eBook)
DOI 10.1007/978-3-540-38505-9

Originally published by Springer-Verlag Berlin Heidelberg New York in 1981.

2061/3020-543210

PREFACE

This book is intended to be a fairly comprehensive treatment of large-scale interconnected systems from an input-output viewpoint. Prior to treating the question of stability (and instability), we study both the decomposition and the well-posedness of such systems. It is not necessary for the reader to have studied feedback stability before tackling this book, as we develop results concerning feedback systems as special cases of more general results pertaining to large-scale systems. However, the reader should know some elementary functional analysis (e.g. Lebesgue spaces, contraction mapping theorem), and have some general knowledge (e.g. Perron-frobenius theorem). The first chapter is introductory, and chapters 2 and 3 contain background material; after that, the remaining chapters are essentially independent and can be read in any order.

I thank Peter Moylan for his careful reading of the manuscript and for several constructive suggestions, and my wife Shakunthala for her support. Virtually all of my research reported in this book was carried out, and most of the book was written, while I was employed by Concordia University, Montreal. I would like to acknowledge research support from the Natural Sciences and Engineering Research Council of Canada, and to a lesser extent from the U.S. Department of Energy. Finally, my thanks to Monica Etwaroo and Jane Skinner for typing the manuscript.

Waterloo
September 29, 1980

M. Vidyasagar

TABLE OF CONTENTS

TABLE OF CONTENTS CONT'D.....

CHAPTER 1: INTRODUCTION

During the past decade or so, there has been a great deal of interest in the study of large-scale systems as a separate discipline in itself. This interest is traceable to many factors, including the growing realization that many physical systems (e.g. power networks, large-scale integrated circuits, etc.) can in fact be viewed as interconnections of several simpler subsystems, and that much valuable information and "structure" is lost if the method of analysis does not take into account the interconnected nature of the system at hand. Moreover, several subjects dealing with "small" systems have reached maturity, so that in order to expand the horizons of knowledge by tackling new and challenging problem areas, researchers have set their sights on large-scale systems. Some prime examples of this are optimal control theory, and the stability theory of single-loop feedback systems.

It is as yet too soon to claim that there exists a comprehensive theory of large-scale systems. Nevertheless, the stability theory of large-scale systems is a well-developed area in which a large variety of results is available. There are in effect two methodologies in stability theory, namely Lyapunov methods and input-output methods. While there are some connections between Lyapunov stability and input-output stability, the actual techniques used to establish the two types of stability are rather different; this is especially so in the case of large-scale systems. Lyapunov methods for large-scale systems are well-documented in the recent books by Michel and Miller [Mic. 1] and Siljak [Sil. 1]. However, though [Mic. 1] contains come input-output results, there is not at present a comprehensive book on the input-output analysis of large-scale systems. In the same vein, the books by Willems [Wil. 2] and Desoer and Vidyasagar [Des. 2] cover feedback systems quite thoroughly from an input-output viewpoint, and it is natural to attempt a similar treatment of large-scale systems.

This book is intended to be a high-level research monograph that sets forth most of the available results on the decomposition, well-posedness, stability and instability of large-scale systems, that can be obtained by input-output methods. Since many results for feedback systems can be obtained as special cases of those given here for large-scale systems, it is not necessary to have read [Wil. 2] or [Des. 2] to follow this book. Though the emphasis here is on input-output stability, we note that input-output methods can be used to establish Lyapunov stability as well. In particular, if a nonlinear system is input-output stable, reachable and detectable, then it is also globally asymptotically stable in the sense of Lyapunov (see [Wil. 3], [Moy. 4]).

Throughout this book, the emphasis is on treating the large-scale system at hand as an <u>interconnected</u> system, consisting of several <u>subsystems</u> interacting through various <u>interconnection operators</u>. (For a precise description, see Section 2.2). It is of course possible to "aggregate" the various subsystem operators and the various interconnection operators, so that the large-scale system at hand is recast in the form of a "single-loop" feedback system. With this reformulation, all of the standard single-loop feedback stability results, such as those in [Des. 2] and [Wil. 2] become applicable. Therefore, whether a given system is a "single-loop" system or an "interconnected" system depends on the method of analysis used to tackle it. However, it can be easily shown that converting the system into a "single-loop" formulation gives unnecessarily conservative stability criteria and well-posedness criteria.

Therefore, in this book we only present results that pertain to interconnected systems, whereby the analysis is carried out in terms of the subsystem operators and the interconnection operators; we avoid treating the system as a whole. For this reason, we exclude linear time-invariant systems from our study. The reason is that, though one can derive necessary and sufficient conditions for the stability and well-posedness of interconnected linear time-invariant systems, the necessary conditions (of necessity) involve tackling the system as a whole. A subsystem level analysis can produce sufficient conditions for stability and sufficient conditions for instability, but not necessary and sufficient conditions.

The book is organized as follows: In Chapter 2, we introduce the concepts of truncations and extended spaces, which provide the mathematical setting for input-output analysis, we then give precise definitions of well-posedness and stability. In Chapter 3, we introduce the concepts of gain and dissipativity, which play an important role in the various criteria for stability and instability, and give explicit methods for computing gains and testing dissipativity.

In Chapter 4, we present a few graph-theoretic techniques for the efficient decomposition of large-scale interconnected systems. Specifically, we show that by identifying the so-called strongly connected components (SCC's) of a given system, we can determine the well-posedness and stability of the original system by studying only the SCC's. In Chapter 5, we present some sufficient conditions for the well-posedness of a system. These criteria are graph-theoretic in nature and can be given a very nice physical interpretation.

In Chapter 6, we give some generalizations of the single-loop "small gain" theorem to arbitrary interconnected systems, while in Chapter 7, we state and prove several generalizations of the single-loop "passivity" theorem. In Chapter 8, we derive several L_2-instability criteria for large-scale systems. Finally, in Chapter 9, we show how the technique of exponential weighting can be used to study L_∞-stability and L_∞-instability using the results of Chapters 6 to 8.

CHAPTER 2: MATHEMATICAL PRELIMINARIES

2.1 TRUNCATIONS, EXTENDED SPACES, CAUSALITY

In this section, we briefly introduce the mathematical notation and terminology that is used throughout this book. The particular notation employed here is from [Vid. 4] and [Des. 2].

Let X denote the set of all real-valued measurable functions mapping R_+ into R, where R denotes the set of real numbers, $R_+ = [0,\infty)$, and the measurability is with respect to the Lebesgue measure. As is customary, we define various subsets of X as follows:

1 Definition For $p \in [1,\infty)$, the set $L_p = L_p[0,\infty)$ denotes the set of all functions $f(.)$ in X such that the function $t \to [|f(t)|]^p$ is integrable over $[0,\infty)$. In other words, $f(.) \in L_p$ for a fixed $p \in [1,\infty)$ if and only if

2 $$\int_0^\infty |f(t)|^p \, dt < \infty$$

Similarly, $L_\infty = L_\infty[0,\infty)$ denotes the set of all functions $f(.)$ in X such that $f(.)$ is essentially bounded over the interval $[0,\infty)$. If $p \in [1,\infty)$, we define the function $||.||_p : L_p \to R_+$ by

3 $$||f||_p = \left[\int_0^\infty |f(t)|^p \, dt \right]^{1/p} , \quad \forall f \in L_p$$

If $p = \infty$, we define $||.||_\infty : L_\infty \to R_+$ by

4 $$||f||_\infty = \operatorname*{ess.\ sup.}_{t \in [0,\infty)} |f(t)|$$

$$= \inf \{ r : \mu[t: |f(t)| > r] = 0 \} , \quad \forall f \in L_\infty$$

where $\mu[.]$ denotes the Lebesgue measure of a set.

It is well-known [Dun. 1, p. 146] that for each $p \in [1,\infty]$, the ordered pair $(L_p , ||.||_p)$ constitutes a Banach space.

In order to have a mathematical framework whereby we can study "unstable" as well as "stable" systems, we introduce the concept of truncated functions and extended spaces.

5 Definition Let $T < \infty$; then the operator $P_T : X \to X$ is defined by setting

6 $$(P_T x)(t) = \begin{cases} x(t) & t \in [0,T] \\ 0 & t > T \end{cases} \qquad \forall x \in X$$

For brevity, we use the notation x_T to denote the function $P_T x$, and refer to $x_T(.)$ as the truncation of a given function $x(.)$ to the interval $[0,T]$.

Note that the operator P_T is a projection on X in the sense that $P_T \cdot P_T = P_T$.

7 Definition For a fixed $p \in [1,\infty]$, the symbol $L_{pe} = L_{pe}[0,\infty)$ denotes the set of all functions $f(.)$ in X such that $f_T(.) \in L_p$ $\forall T < \infty$ (though $f(.)$ itself may or may not belong to L_p). The space L_{pe} is referred to as the extension of the space L_p .

8 Example The function $f_1(t) = t$ belongs to the extended spaces L_{pe} for all $p \in [1,\infty]$, but does not belong to any of the unextended spaces L_p for $p \in [1,\infty]$. The function $f_2(t) = \tan t$ does not belong to any of the unextended spaces L_{pe} .

It is clear that, for each fixed p, we have $L_p \subset L_{pe} \subset X$. Moreover, since $L_p\,[0,T]$ is a subset of $L_1\,[0,T]$ for all finite T , it is clear that $L_{pe} \subset L_{1e}$ $\forall p \in [1,\infty]$. Thus L_{1e} is the largest set that we need to use in this book.

9 Definition Let $p \in [1,\infty]$ be fixed, and let $T < \infty$. Then for every $f \in L_{pe}$, the truncated norm $||f||_{Tp}$ is defined by

10 $$||f||_{Tp} = ||f_T||_p = ||P_T f||_p$$

Let $p = 2$, and let $T < \infty$. Then for every $f, g \in L_{2e}$, the truncated inner product $\langle f,g\rangle_T$ is defined by

11 $$\langle f,g\rangle_T = \langle f_T, g_T\rangle = \int_0^T f(t)\, g(t)\, dt$$

Note that, for every $p \in [1,\infty]$ and every $f \in L_{pe}$, the quantity $||f||_{Tp}$ is a well-defined finite real number for every $T < \infty$, though $||f||_p$ is defined only if f actually belongs to the unextended space L_p. Moreover, we have

12 Lemma Let $p \in [1,\infty]$ and let $f \in L_p$. Then $||f||_{Tp}$ is a nondecreasing function of T. Furthermore, $f \in L_p$ (the unextended space) if and only if there exists a finite constant c such that $||f||_{Tp} \le c \quad \forall T < \infty$. In this case, $||f||_{Tp} \to ||f||_p$ as $T \to \infty$, monotonically from below.

The proof is obvious and is left to the reader.

In order to deal with systems having multiple inputs and outputs, we introduce the spaces L_p^n and L_{pe}^n.

13 Definition Let $p \in [1,\infty]$ and let $n \ge 1$ be an integer. Then the set L_p^n (respectively L_{pe}^n) consists of all n-tuples $\underset{\sim}{f}(.) = [f_1(.)\ f_2(.)\ \ldots\ f_n(.)]'$, where $f_i(.) \in L_p$ (respectively L_{pe}) $\forall i$. The norm of a function $\underset{\sim}{f}(.) \in L_p^n$ is is defined by

14 $$||\underset{\sim}{f}(.)||_p = \begin{cases} \{\int_0^\infty ||f(t)||^p\, dt\}^{1/p} & \text{if } p < \infty, \\ \underset{t \in [0,\infty)}{\text{ess. sup}} ||f(t)|| & \text{if } p = \infty \end{cases}$$

where $||.||$ is the Euclidean norm on R^n.

Note that we use the same symbol $||.||_p$ to denote both the norm on L_p as well as the norm on L_p^n. This usage is not expected to cause any confusion.

In the definition of the norm $||.||_p$ on L_p^n, the choice of the vector norm $||.||$ on R^n is essentially arbitrary. However, with the choice in (14), the space L_2^n is an inner product space, and the inner product $\langle \underset{\sim}{f},\underset{\sim}{g}\rangle$ of two elements $\underset{\sim}{f},\underset{\sim}{g}$ in L_2^n is defined by

15 $$\langle f,g\rangle = \int_0^\infty \underset{\sim}{f}'(t)\, \underset{\sim}{g}(t)\, dt = \sum_{i=1}^{n} \langle f_i, g_i\rangle$$

where $f_i(.)$ and $g_i(.)$ are the component functions of $\underset{\sim}{f}(.)$ and $\underset{\sim}{g}(.)$, respectively. The truncated norm $||.||_T : L^n_{pe} \to R_+$ and the truncated inner product $\langle .,.\rangle_T : L^n_{2e} \times L^n_{2e} \to R$ are defined in a manner analogous to (10) and (11), respectively.

To study discrete-time systems, we introduce the space of sequences S and its various subsets.

16 <u>Definition</u> The set S consists of all sequences $\{x^{(i)}\}_{i=0}^{\infty}$. For $1 \leq p < \infty$, the set ℓ_p consists of all $\{x^{(i)}\}$ in S such that

17 $$\sum_{i=0}^{\infty} |x^{(i)}|^p < \infty$$

The set ℓ_∞ consists of all bounded sequences in S. For $p \in [1,\infty)$, we define the function $||.||_p : \ell_p \to R_+$ by

18 $$||x||_p = \left(\sum_{i=0}^{\infty} |x^{(i)}|^p \right)^{1/p}$$

We also define $||.||_\infty : \ell_\infty \to R_+$ by

19 $$||x||_\infty = \sup_i |x^{(i)}|$$

We can also define the concept of truncations in the present context.

20 <u>Definition</u> For each $i \geq 0$, the operator $P_i : S \to S$ is defined by

21 $$(P_i x)^{(j)} = \begin{cases} x^{(j)} & 0 \leq j \leq i \\ 0 & j > i \end{cases}$$

Finally, we define the spaces S^n and ℓ^n_p.

22 <u>Definition</u> Let n be a positive integer. Then the set S^n (resp. ℓ^n_p) is defined as the set of all sequences of

n-tuples $\{\underset{\sim}{x}^{(i)}\} = [x_1^{(i)}, x_2^{(i)}, \ldots, x_n^{(i)}]'$ such that $\{x_j^{(i)}\} \in S$ (resp. ℓ_p) $\forall j$. The norm $||.||_p : \ell_p^n \to R_+$ is defined by

23 $$||x||_p = \begin{cases} \left(\sum_{i=1}^{\infty} ||x^{(i)}||^p\right)^{1/p} & \text{if } p < \infty \\ \sup_i ||x^{(i)}|| & \text{if } p = \infty \end{cases}$$

where $||.||$ denotes the Euclidean norm on R^n .

We next introduce the concept of causality.

24 Definition An operator $G : L_{1e} \to L_{1e}$ is said to be causal if

25 $$P_T G = P_T G P_T \quad , \quad \forall T < \infty$$

or, equivalently,

26 $$(Gf)_T = (Gf_T)_T \; , \quad \forall T < \infty \; , \quad \forall f \in L_{1e}$$

27 Lemma An operator $G : L_{1e} \to L_{1e}$ is causal in the sense of Definition (24) if and only if the following is true : whenever f and g are two functions in L_{1e} such that $f_T = g_T$ for some $T < \infty$, we have $(Gf)_T = (Gg)_T$.

Proof For the sake of clarity, let us say that an operator $G : L_{1e} \to L_{1e}$ has property (s) if

28 $$f_T = g_T \Rightarrow (Gf)_T = (Gg)_T$$

We must show that causality in the sense of Definition (24) is equivalent to property (s) . To show this, suppose first that G is causal, and that $f_T = g_T$ for some T . Then by (25), we have

29 $$(Gf)_T = (Gf_T)_T = (Gg_T)_T = (Gg)_T$$

so that G has property (s) . Conversely, suppose G has property (s) . Since $f_T = (f_T)_T$ $\forall T$, we have from (28) that

30 $$(Gf)_T = (Gf_T)_T$$

so that G is causal. □

It is clear that there is nothing special about operators on L_{1e} as far as causality goes, and that one can equally well define causality with respect to operators from L^n_{pe} to L^m_{qe}, or from S^n to S^m, where $p, q \in [1,\infty]$ and n,m are positive integers.

We conclude this section by introducing the set A, which plays an important role in the study of linear time-invariant operators.

31 Definition The set A consists of all distributions $f(.)$ of the form

32 $$f(t) = \begin{cases} 0, & t < 0 \\ \sum_{i=0}^{\infty} f_i \; \delta(t-t_i) + f_a(t) , & t \geq 0 \end{cases}$$

where $\delta(.)$ denotes the unit impulse distribution, $0 \leq t_0 < t_1 < \ldots$ are real constants, and $\{f_i\} \in \ell_1$, $f_a(.) \in L_1$. The norm $||.||_A$ on A is defined by

33 $$||f(.)||_A = \sum_{i=0}^{\infty} |f_i| + \int_0^{\infty} |f_a(t)| \, dt$$

The product of two elements $f(.)$ and $g(.)$ in A is defined as their convolution; i.e.,

34 $$(f*g)\;(t) = \int_0^t f(t-\tau)\; g(\tau)\; d\tau = \int_0^t f(\tau)\; g(t-\tau)\; d\tau$$

Remarks Basically, the set A consists of L_1 augmented by delayed impulses. It is easy to see that L_1 is a subset of A, and that if $f(.) \in L_1$, then $||f(.)||_1 = ||f(.)||_A$. Moreover, the ordered pair $(A, ||.||_A)$ is a Banach space.

In (34), one should interpret

35 $$\delta(t-t_a) * \delta(t-t_b) = \delta(t-t_a-t_b)$$

36 $$\delta(t-t_a) * f_a(t) = f_a(t-t_a)$$

Thus, if f and g are of the form

37 $$f(t) = \sum_{i=0}^{\infty} f_i \, \delta(t-t_i) + f_a(t)$$

38 $$g(t) = \sum_{i=0}^{\infty} g_i \, \delta(t-\tau_i) + g_a(t)$$

then

39 $$(f*g)(t) = \sum_{i=0}^{\infty} \sum_{j=0}^{\infty} f_i \, g_j \, \delta(t-t_i-\tau_j) + \sum_{i=0}^{\infty} f_i \, g_a \, (t-t_i) + \sum_{j=0}^{\infty} g_j \, f_a(t-\tau_j) + \int_0^t f_a(t-\tau) \, g_a(\tau) \, d\tau$$

It is routine to verify from (39) that convolution is commutative, left- and right- distributive with respect to addition, and that

40 $$||f*g||_A \leq ||f||_A \cdot ||g||_A$$

Also, we see from (39) that

41 $$f*\delta = \delta*f = f \, , \quad \forall f \in A$$

Hence the set A is a Banach algebra with a unit, with $||.||_A$ as the norm, $*$ as the product, and δ as the unit.

Given any $f(.) \in A$, the integral

42 $$\hat{f}(s) = \int_0^{\infty} f(t) \, e^{-st} \, dt$$

is well-defined whenever $\text{Re } s \geq 0$, and in fact,

43 $$|\hat{f}(s)| \leq ||f(.)||_A \, , \quad \forall s \in C_+$$

where $C_+ = \{s: \text{Re } s \geq 0\}$. Thus every element $f(.)$ of A is Laplace transformable, and the region of convergence of the

Laplace transform $\hat{f}(.)$ includes the closed right half-plane C_+. For notational convenience, we introduce the set $\hat{A}$.

44 **Definition** The set $\hat{A}$ consists of the Laplace transforms of the elements of A.

Since convolution in the time domain is equivalent to pointwise multiplication in the s-domain, we see that sums and products of elements of $\hat{A}$ once again belong to $\hat{A}$. Also, it can be shown quite easily that any $\hat{f} \in \hat{A}$ is continuous at every $s \in C_+$, and analytic at every $s \in C_+^o$ (where $C_+^o = \{s: \text{Re } s > 0\}$ denotes the interior of C_+). Finally, (43) shows that every element of $\hat{A}$ is bounded over C_+.

Once we have defined A and $\hat{A}$, we can also define $A^{n\times m}$ and $\hat{A}^{n\times m}$ in an obvious way. We next define the extension of A, denoted by A_e.

45 **Definition** The set A_e consists of all distributions $f(.)$ such that $f_T(.) \in A, \; \forall T \geq 0$.

Note that Definition (45) is entirely analogous to Definition (7).

The set A_e is important because it can be shown that, if G is a linear convolution operator of the type

46 $$(Gf)(t) = \int_0^\infty g(t-\tau)\; f(\tau)\; d\tau$$

then G is causal and maps L_{pe} into itself $\forall p \in [1,\infty]$, if and only if the kernel (or "impulse response") $g(.) \in A_e$. The proof of this important fact can be obtained by adapting that of [Des. 2, Theorem IV.7.5]. Thus, all linear convolution operators that we encounter in this monograph (even the "unstable" ones) can be assumed to be of the form (46), where the kernel $g(.) \in A_e$ (or, more generally, $g(.) \in A_e^{n\times m}$, in the case of a multivariable system. Similarly, it can be shown [Vid. 4, Thrm.6.5.37] that, if G is of the form (46), then G maps L_p (the unextended space) into itself $\forall p \in [1,\infty]$, if and only if $g(.) \in A$. This shows that the set A essentially consists of all "stable" impulse responses (see Definition 3.1.1).

2.2 DEFINITIONS OF WELL-POSEDNESS AND STABILITY

In this section, we delineate the class of large-scale interconnected systems under study in this book, and we give precise definitions of what is meant by such a system being well-posed or stable.

Throughout this book, we shall be concerned with analysis of a large-scale interconnected system (LSIS) described by the set of equations

1a $$e_i = u_i - \sum_{j=1}^{m} H_{ij}\, y_j$$

1b $$y_i = G_i\, e_i \qquad\qquad i = 1,\ldots,m$$

where u_i, e_i, y_i all belong to the extended space $L_{pe}^{n_i}$ for a fixed $p \in [1,\infty]$ and some positive integer n_i, the operator G_i maps $L_{pe}^{n_i}$ into itself, and the operator H_{ij} maps $L_{pe}^{n_j}$ into $L_{pe}^{n_i}$.* We can refer to u_i, e_i, y_i as the i-th input, error, and output, respectively. Where convenient, we use the symbol u to denote the m-tuple $(u_1, \ldots, u_m)$, e to denote $(e_1, \ldots, e_m)$, and y to denote $(y_1, \ldots, y_m)$. Note that u, e, y all belong to the product space L_{pe}^{n}, where $n = \sum_{i=1}^{m} n_i$. In the same spirit, we sometimes use the symbols G and H to denote operators from L_{pe}^{n} into itself defined by

2 $$G = \begin{bmatrix} G_1 & & 0 \\ & \ddots & \\ 0 & & G_m \end{bmatrix}$$

---

*To avoid a proliferation of symbols, we assume that the system G_i has an equal number of inputs and outputs. This assumption is entirely dispensable.

3 $$H = \begin{bmatrix} H_{11} & \cdots & H_{1m} \\ \vdots & & \vdots \\ H_{m1} & \cdots & H_{mm} \end{bmatrix}$$

With these definitions, the system equations (1) can be compactly expressed as

4a $$e = u - Hy$$

4b $$y = Ge$$

The system description (1) is quite general and is capable of representing several types of physical systems. One can think of (1) as representing several "isolated" or "decoupled" subsystems, corresponding to the operators $G_1,\ldots,G_m$, such that the input to G_i is a linear combination of an external input u_i and several "interaction" signals $H_{ij}\, y_j$. This is depicted in Figure 2.1 .

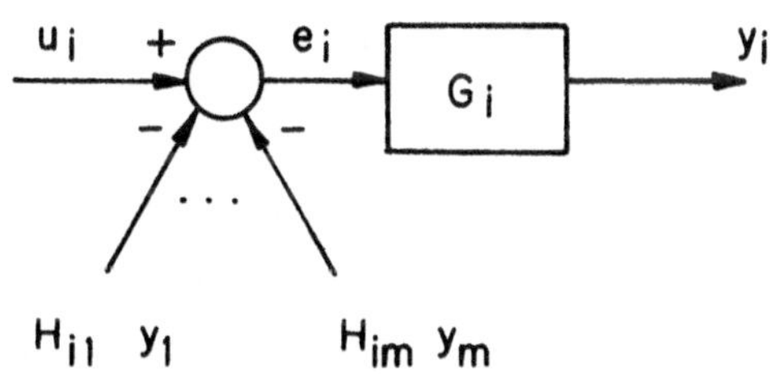

FIGURE 2.1

For this reason, we refer to m as the number of subsystems, $G_1,\ldots,G_m$ as the <u>subsystem operators,</u> and $H_{11},\ldots,H_{mm}$ as the <u>interconnection operators.</u>

In some cases, particularly in proving dissipativity-type theorems for stability and instability, (Chapters 7 and 8) we assume that for all i,j, the interconnection operator H_{ij}: $L_{pe}^{n_j} \to L_{pe}^{n_i}$ can be represented by an $n_i \times n_j$ matrix $\underset{\sim}{H}_{ij}$ of constant real numbers, i.e. that

5 $$(H_{ij}\, y_j)(t) = \underset{\sim}{H}_{ij}\, y_j(t) \;, \quad \forall t, \quad \forall y_j \in L_{pe}^{n_j}$$

Actually, this assumption does not result in any loss of generality, because this assumption can always be satisfied by increasing the number of subsystems (m) if necessary. (If a particular operator H_{ij} cannot be represented by a constant matrix, then increase m by one and include H_{ij} among the operators G_i). If all interconnection operators can be represented by constant matrices, then we refer to the constant n×n matrix $\underset{\sim}{H}$ defined by

6 $$\underset{\sim}{H} = \begin{bmatrix} \underset{\sim}{H}_{11} & \cdots & \underset{\sim}{H}_{1m} \\ \vdots & & \vdots \\ \underset{\sim}{H}_{m1} & & \underset{\sim}{H}_{mm} \end{bmatrix}$$

as the <u>interconnection matrix</u>.

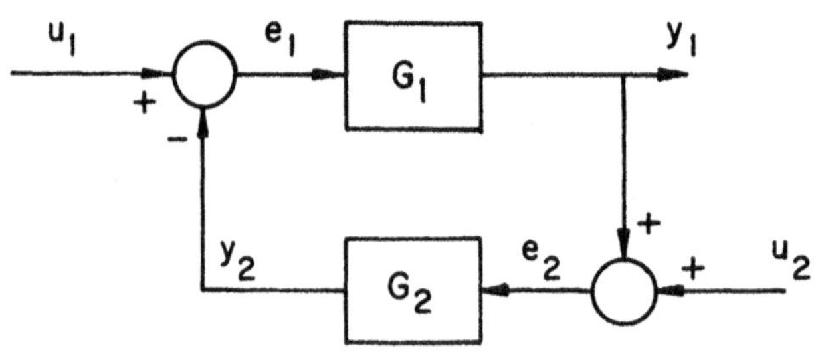

FIGURE 2.2

The standard <u>feedback</u> configuration, shown in Figure 2.2 and studied in detail in [Des. 1] and [Wil. 1] among others, is a special case of the system description (1) . The feedback system of Figure 2.2 is described by

7a $$e_1 = u_1 - y_2$$

7b $$e_2 = u_2 + y_1$$

7c $$y_1 = G_1 e_1$$

7d $$y_2 = G_2 e_2$$

where $u_1, u_2, e_1, e_2, y_1, y_2$ all belong to L_{pe}^{ν} for some fixed $p \in [1,\infty]$ and some positive integer ν, and G_1, G_2 map L_{pe}^{ν} into itself.* To put the system (7) in the form (1), let $m = 2$ (two subsystems), $n_1 = n_2 = \nu$, $n = 2\nu$, G_1, G_2 as in (7), and

8 $$\underset{\sim}{H} = \begin{bmatrix} \underset{\sim}{0}_{\nu} & \underset{\sim}{I}_{\nu} \\ -\underset{\sim}{I}_{\nu} & \underset{\sim}{0}_{\nu} \end{bmatrix}$$

where $\underset{\sim}{0}_{\nu}$ and $\underset{\sim}{I}_{\nu}$ denote the null matrix and identity matrix of order $\nu\times\nu$ respectively. Note that the inter-connection operators can be represented by constant matrices in this case, and that the interconnection matrix $\underset{\sim}{H}$ is skew-symmetric and invertible. These properties are implicity used in much of feedback stability theory.

Comparing the general large-scale system description (1) with the feedback system description (7), we see that if we <u>aggregate</u> the equations (1) into the form (4), then (4) and (7) are very similar. In fact, (4) is a special case of (7), with $u_1 = u$, $u_2 = 0$, $G_1 = G$, $G_2 = H$, $e_1 = e$, and $y_1 = y$. This is shown in Figure 2.3. Thus, given an LSIS, one can either represent it in the <u>decomposed</u> form (1) and analyze it at the <u>subsystem level</u>, or one can represent it in the <u>aggregated form</u> (4) and analyze it as a <u>single-loop system</u>. If one chooses the latter option, one can immediately apply all of the standard results derived in [Des. 2] and [Wil. 2] for feedback systems. Thus the main emphasis in this monograph is on analyzing a given LSIS at the subsystem level, taking full advantage of the fact that the system at hand is an interconnection of several (presumably simpler) subsystems.

---

*Actually, u_1, e_1, y_2 all need to belong to the same space $L_{pe}^{\nu_1}$ and u_2, y_1, e_2 all need to belong to the same space $L_{qe}^{\nu_2}$, but in general we could have $p \neq q$, $\nu_1 \neq \nu_2$. The extension of the results presented here to this situation is transparent.

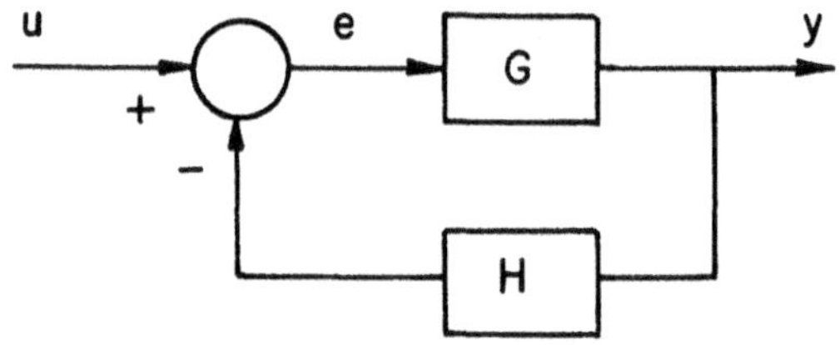

FIGURE 2.3

With regard to the system described by the set of equations (1) (or, equivalently, (4)), one can ask basically two types of questions. The first type of question takes the following form: Does the system (1) have a unique set of solutions e,y in L^n_{pe} corresponding to each set of inputs $u \in L^n_{pe}$? If so, is the dependence of e,y on u causal, and globally Lipschitz continuous? The definition and study of the <u>well-posedness</u> of the system (1) takes into account such considerations. The second type of question takes the following form: Given a set of inputs $u \in L^n_p$ (the <u>unextended</u> space) and <u>assuming</u> that the system equations (1) have one or more solutions for e,y in L^n_{pe}, do these solutions in fact belong to L^n_p ? If so, does the relation mapping u into (e,y) have "finite gain"? The definition and study of the <u>stability</u> of the system (1) takes into account such considerations as the above. The reason for separating the two types of questions is that usually the conditions that imply well-posedness are quite distinct and different in nature from the conditions that imply stability. This can be seen by comparing Chapter 5 with Chapters 6 to 9 .

We now turn to the definitions.

9 <u>Definition</u> The system (1) is said to be <u>well-posed</u> if the following conditions hold:

(W1) For each set of inputs $u \in L^n_{pe}$, there exists a unique set of errors $e \in L^n_{pe}$ and a set of outputs $y \in L^n_{pe}$ such that the system equations (1) are satisfied.

(W2) The dependence of e and y on u is causal; i.e. whenever $u^{(1)}$ and $u^{(2)}$ are two input sets in L^n_{pe} such that for some $T > 0$ we have

10 $$u_T^{(1)} = u_T^{(2)}$$

then the corresponding solution sets $\{e^{(1)}, y^{(1)}\}$ and $\{e^{(2)}, y^{(2)}\}$ satisfy

11 $$e_T^{(1)} = e_T^{(2)}$$

12 $$y_T^{(1)} = y_T^{(2)}$$

(W3) For each finite T, the dependence of e_T and y_T on u_T is globally Lipschitz continuous. In other words, for each $T < \infty$, there exists a finite constant k_T such that, whenever $u^{(1)}$ and $u^{(2)}$ are two sets of inputs in L^n_{pe} and $\{e^{(1)}, y^{(1)}\}$, $\{e^{(2)}, y^{(2)}\}$ are the corresponding solution sets of (1), we have

13 $$||e^{(1)}-e^{(2)}||_{Tp} \leq k_T||u^{(1)}-u^{(2)}||_{Tp}$$

14 $$||y^{(1)}-y^{(2)}||_{Tp} \leq k_T||u^{(1)}-u^{(2)}||_{Tp}$$

The above definition of well-posedness is quite broad, as it implies (i) existence and uniqueness of solutions to the system equations, (ii) causal dependence of solutions on inputs, and (iii) global Lipschitz continuity of solutions as functions of the input. Note that (W2) is actually implied by (W3); however, we list (W2) as a separate condition in order to highlight it. The definition of well-posedness in [Wil.2] requires something more: namely that all of the above properties be preserved when G_i, H_{ij} are perturbed slightly. We do not make this a part of the definition of well-posedness. However, it is shown in Chapter 5 that properties (W1) - (W3) are preserved if each G_i, H_{ij} is replaced by another operator of the same or higher "class" .

Notice that in adopting the system description (1), we assume that each subsystem is represented by an operator, instead of a multi-valued relation. Now, suppose for instance that the output y_i is obtained from the error e_i by solving a set of differential equations. The fact that $G_i : e_i \mapsto y_i$ is an operator means that this set of differential equations is well-posed in another sense. Definition (9) is not concerned with such questions. Rather, we assume that each subsystem and interconnection is represented by an operator, and ask about the well-posedness of the overall system from an input-output point of view.

It is important to note that a system that is well-posed in the sense of Definition (9) is precluded from having finite escape time because $u \in L_{pe}^n$ implies $e,y \in L_{pe}^n$. Moreover, in a well-posed system, the maps $P_T u \mapsto P_T e$ and $P_T u \mapsto P_T y$ are Lipschitz continuous maps from $L_p^n [0,T]$ into itself, for all finite T, and are hence stable in the sense of Definition (15) below. However, this does not mean that the maps $u \mapsto e$ and $u \mapsto y$ are stable in the sense of Definition (15).

Having defined the well-posedness of the system (1), we now turn our attention to stability.

15 Definition The system (1) is said to be L_p-stable (or merely stable if p is obvious from the context) if the following properties hold:

(S1) For each set of inputs $u \in L_p^n$, we have that any e,y in L_{pe}^n such that (1) is satisfied, actually belong to L_p^n .

(S2) There exist finite constants γ_p and b_p such that, whenever $u \in L_p^n$ and $e,y \in L_p^n$ are some corresponding solutions of (1), we have

16 $$||e||_p \leq \gamma_p ||u||_p + b_p$$

17 $$||y||_p \leq \gamma_p ||u||_p + b_p$$

If $b_p = 0$, we say that the system (1) is <u>L_p-stable with zero bias</u>.

It is important to note that in Definition (15), we do <u>not</u> assume existence and/or uniqueness of solutions to (1). If for a particular $u \in L_p^n$, no $e, y \in L_{pe}^n$ exist such that (1) is satisfied, then conditions (S1) and (S2) are satisfied vacuously. In this way, the stability issue is divorced from the issue of well-posedness. Also, note that the system (1) is L_p-stable if and only if the relation mapping u into (e,y) has finite gain, in the sense of Definition (3.1.1). Similarly, the system (1) is L_p-stable with zero bias if and only if the relation mapping u into (e,y) has finite gain with zero bias, the sense of Definition (3.1.1).

Note that, in order for the system (1) to be <u>L_p-unstable</u> in the sense of Definition (15), one of two things must happen : either (i) there exist a set of inputs $u \in L_p^n$ and a set of errors/outputs $e, y \in L_p^n$ such that (1) is satisfied, but either e or y does not belong to L_p^n, or (ii) there exists a sequence of input sets $u^{(j)} \in L_p^n$ and a corresponding sequence of error/output sets $e^{(j)} \in L_p^n$, $y^{(j)} \in L_p^n$ such that one of the sequence $\{||e^{(j)}||_p / ||u^{(j)}||_p\}$ or $\{||y^{(j)}||_p / ||u^{(j)}||_p\}$ is unbounded. (In this case, the system (1) may still have a form of "stability" in that $u \in L_p^n \Rightarrow e \in L_p^n$, $y \in L_p^n$; however, it does not satisfy (16) - (17)).

We now present and discuss an alternative formulation of the system description and stability definition. It is clear that, in the system description (1), there are two types of operators, namely the subsystem operators (G_i) and the interconnection operators (H_{ij}). This distinction serves a useful purpose in deriving the results of Chapters 6-9, where the kinds of conditions imposed on the subsystem operators are quite different in nature from those imposed on the interconnection operators. However, in Chapter 4 and 5, where we apply some techniques from graph theory to study the decomposition and well-posedness of large-scale interconnected systems, the distinction between subsystem and interconnection operators disappears entirely. In these situations, it is therefore logical to adopt a system description that also makes no distinction between subsystem and interconnection operators.

To meet this objective, in Chapter 4 and 5 we study systems described by

18 $$e_i = u_i - \sum_{j=1}^{m} S_{ij}\, e_j \ , \quad i=1,\ldots,m$$

where $e_i,\ u_i \in L_{pe}^{n_i}$ for some fixed p and some positive integer n_i, and $S_{ij} : L_{pe}^{n_j} \to L_{pe}^{n_i}$. We see that this system consists of m summing junctions, where the inputs to each summing junction consist of an external input (u_i) and interaction signals from other summing junction outputs $(S_{ij}\ e_j)$. This depicted in Figure 2.4

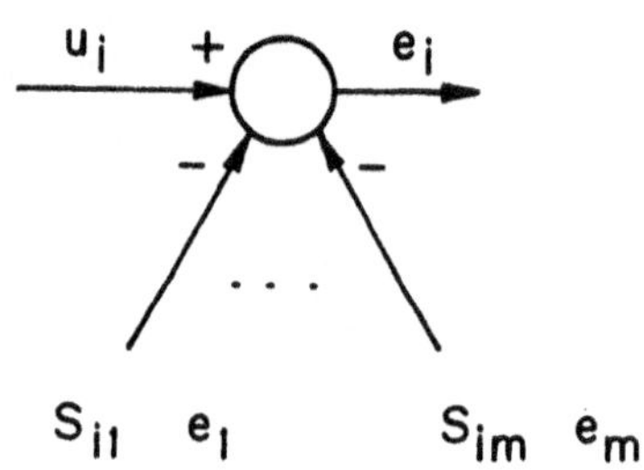

FIGURE 2.4

19 <u>Definition</u> The system (18) is said to be <u>well-posed</u> if the following conditions hold:

(W1) For each set of inputs $u \in L_{pe}^{n}$, there exists a unique set of errors $e \in L_{pe}^{n}$ such that the system equations (18) are satisfied.

(W2) The dependence of e and $\{S_{ij}\ e_j\}$ on u is causal; i.e., whenever $u^{(1)}$ and $u^{(2)}$ are two input sets in L_p^n such that for some $T > 0$ we have

20 $$u_T^{(1)} = u_T^{(2)}$$

then the corresponding solution sets $e^{(1)}$ and $e^{(2)}$ of (18) satisfy

21 $$e_T^{(1)} = e_T^{(2)}$$

22 $$(S_{ij}\ e_j^{(1)})_T = (S_{ij}\ e_j^{(2)})_T\ ,\quad \forall i,j$$

(W3) For each finite T, the dependence of e_T and $S_{ij}\ e_{jT}$ on u_T is globally Lipschitz continuous. In other words, for each $T < \infty$ there exists a finite constant k_T such that, whenever $u^{(1)}$ and $u^{(2)}$ are two sets of inputs in L_{pe}^n and $e^{(1)}$, $e^{(2)}$ are the corresponding solution sets of (18), we have

23a $$||e^{(1)}-e^{(2)}||_{Tp} \le k_T||u^{(1)}-u^{(2)}||_{Tp}$$

23b $$||S_{ij}\ e_j^{(i)} - S_{ij}\ e_j^{(2)}||_{Tp} \le k_T||u^{(1)}-u^{(2)}||_{Tp}$$

24 Definition The system (18) is said to be L_p-stable if the following properties hold:

(S1) For each set of inputs $u \in L_p^n$, we have that any e in L_{pe}^n such that (18) is satisfied, actually belong to L_p^n, and moreover, we also have that $S_{ij}\ e_j \in L_p^{n_i}\ \forall i,j$.

(S2) There exist finite constants γ_p and b_p such that, whenever $u \in L_p^n$ and $e \in L_p^n$ is some corresponding solution of (18), we have

25 $$||e||_p \le \gamma_p||u||_p + b_p$$

26 $$||S_{ij}\ e_j||_p \le \gamma_p||u||_p + b_p\ ,\quad \forall i,j\ .$$

If $b = 0$, we say that the system (18) is L_p-stable with zero bias.

In Definition (24), we not only require that the explicitly displayed unknowns e_i belong to $L_p^{n_i}$ $\forall i$ whenever the input set $u \in L_p^n$, but also that $S_{ij} e_j \in L_p^{n_i}$ $\forall i,j$ whenever $u \in L_p^n$. Clearly if $e_i \in L_p^{n_i}$ $\forall i$ whenever $u \in L_p^n$, then $\sum_{j=1}^{m} S_{ij} e_j \in L_p^{n_i}$ $\forall i$ whenever $u \in L_p^n$. But Definition (24) requires something more, namely that each of the terms $S_{ij} e_j$ individually belong to $L_p^{n_i}$ whenever $u \in L_p^n$.

The alternative system descriptions (1) and (18), together with the associated definitions of well-posedness and stability, are closely related. To convert a system described by (18) to the form (1), we define

27 $$y_{ij} = S_{ij} e_j , \quad \forall i,j$$

Then (19) can be recast as

28a $$e_i = u_i - \sum_{j=1}^{m} y_{ij}$$

28b $$y_{ij} = S_{ij} e_j$$

which is of the form (1). Moreover, it can be verified that Definition (19) for the well-posedness of the system (28) is equivalent to Definition (9) applied to the system (28). Similarly, Definition (24) for the stability of system (18) is equivalent to Definition (15) applied to the system (28).

To convert a system described by (1) to the form (19), define

29 $$u_i = 0 \quad \text{for} \quad i = m+1,\ldots,2m$$

30 $$e_i = G_{i-m} e_{i-m} \quad \text{for} \quad i = m+1,\ldots,2m$$

Then the system (1) becomes

$$e_i = u_i - \sum_{j=1}^{m} H_{ij}\, e_{j+m} \quad (31a)$$

$$e_{i+m} = u_{i+m} + G_i\, e_i \quad (31b) \qquad \Bigg\}\; i = 1,\ldots,m$$

In order for the system (31) to be well-posed in the sense of Definition (19), we must have the following: (i) Corresponding to each set of inputs $u_1,\ldots,u_{2m}$, (31) must have a unique set of solutions for $e_1,\ldots,e_{2m}$. This condition is more restrictive than the corresponding condition resulting from applying Definition (9) to the system (1), because this latter only requires existence and uniqueness of solutions when $u_{m+1} = \ldots = u_{2m} = 0$. (ii) The quantities $e_1,\ldots,e_{2m}$ as well as $H_{11}e_{m+1},\ldots,H_{mm}e_{2m}$ must depend in a causal manner on $u_1,\ldots,u_{2m}$. If Definition (9) is applied to the system (1), then we only require that $e_1,\ldots,e_{2m}$ depend causally on $u_1,\ldots,u_m$, with no mention of $H_{11}e_{m+1},\ldots,H_{mm}e_{2m}$. (iii) The quantities $P_Te_1,\ldots,P_Te_{2m}$ as well as $P_TH_{11}e_{m+1},\ldots,P_TH_{mm}e_{2m}$ must depend in a globally Lipschitz continuous manner on $P_Tu_1,\ldots,P_Tu_{2m}$. If Definition (9) is applied to the system (1), we only require that quantities $P_Te_1,\ldots,P_Te_{2m}$ depend in a globally Lipschitz continuous manner on $P_Tu_1,\ldots,P_Tu_m$ with no mention of $P_TH_{11}e_{m+1},\ldots,P_TH_{mm}e_{2m}$. Because of all these differences, it is clear that Definition (19) applied to the system (31) gives a stronger (i.e. more restrictive) concept of well-posedness than Definition (9) applied to the system (1). However, if we assume that each of the operators H_{ij} is causal, and that each operator $e_{jT} \mapsto (H_{ij}\, e_j)_T$ has finite incremental gain (see Definition (3.1.1), then the well-posedness of the system (31) in the sense of Definition (19) with $u_{m+1} = \ldots = u_{2m} = 0$, is equivalent to the well-posedness of the system (1) in the sense of Definition (9). An operator H_{ij} satisfying the above conditions is said to be weakly Lipschitz (see Definition (5.1.1)). Thus, our conclusions can be summarized as follows: If H_{ij} is a weakly Lipschitz operator for all i,j, then the system (31) is well-posed with $u_{m+1} = \ldots = u_{2m} = 0$ in the sense of Definition (19) if and only if the system (1) is well-posed in the sense of Definition (9).

Similar remarks apply to the two stability definitions (5) and (24), as applied to the appropriate system descriptions. In order for the system (31) to be L_p-stable in the sense of Definition (24), two conditions must hold: (i) whenever $u_i \in L_p^{n_i}$ for $i = 1,\ldots,2m$, any corresponding $e_1,\ldots,e_{2m}$ must belong to $L_p^{n_1},\ldots,L_p^{n_m}$, respectively; moreover, the relations taking the input set into $(e_1,\ldots,e_{2m})$ must have finite gain (as in (25)). If we restrict u_{m+1} through u_{2m} to be zero, then this is the same condition as is obtained by applying Definition (15) to the system (1). (ii) Whenever $u_i \in L_p^{n_i}$ for $i = 1,\ldots,2m$, any corresponding $H_{11}e_{m+1},\ldots,H_{mm}e_{2m}$ must belong to the appropriate L_p-space; moreover, the relation taking the input set into $(H_{11}e_{m+1},\ldots,H_{mm}e_{2m})$ must have finite gain. This condition does not arise at all when Definition (15) is applied to the system (1). Thus Definition (24) as applied to the system (31) gives a stronger concept of stability than Definition (16) applied to the system (1). However, if all operators H_{ij} have finite gains, then the system (31) is L_p-stable with $u_{m+1} = \ldots u_{2m} = 0$ in the sense of Definition (24) if and only if the system (1) is L_p-stable in the sense of Definition (15).

Now that we have established the interrelationships between the two alternative system descriptions (1) and (19), hereafter we make use of whichever system description is appropriate to the technique being used, without further comment. Specifically, in Chapters 4 and 5, where we study the decomposition and well-posedness, respectively, of large-scale interconnected systems, we use the description (18). The reason is that the results in Chapters 4 and 5 involve the construction of a directed graph associated with the given system, and in this process the distinction between the subsystem operators and the interconnection operators is lost completely. On the other hand, in Chapters 6-9, we use the description (1), because the results there do make a distinction between the subsystem operators and the interconnection operators.

NOTES AND REFERENCES

The use of extended spaces to give a general formulation of the input-output stability problem is due to Sandberg [San. 1,2,3], and Zames [Zam. 1,2,3], who also did pioneering work on feedback stability. Causality is discussed at greater length by Saeks [Sae. 1]. The set A was introduced by Desoer and Wu [Des. 1]. The definition of well-posedness used here differs somewhat from that of Willems [Wil. 2] for feedback systems, but is in the same spirit. The formulation of the stability definition (2.2.24) is due to Callier, Chan and Desoer [Cal. 2].

CHAPTER 3: GAIN AND DISSIPATIVITY

In this chapter, we define various concepts that are used throughout the rest of the book. These include gain, gain with zero bias, incremental gain, passivity, strict passivity, dissipativity, conditional gain and conditional dissipativity. We also discuss how these concepts pertain to the two most commonly encountered types of operators, namely linear integral operators and memoryless nonlinear operators. Since almost all of the stability criteria in the subsequent chapters are couched in terms of the gains or dissipativity constants of various operators, it is important to know how to actually calculate these constants for a given operator. Thus, the results of this chapter are important in order to show that the various stability criteria derived in this monograph can actually be applied to "real" systems.

3.1 GAIN, GAIN WITH ZERO BIAS, AND INCREMENTAL GAIN

3.1.1 Definitions of Various Types of Gain

1 *Definition* Suppose $p \in [1,\infty]$, that n, m are positive integers, and that $G:L^n_{pe} \to L^m_{pe}$ is a given operator. Then we define the *gain* of the operator G, denoted by $\gamma_p(G)$, by

2 $$\gamma_p(G) = \inf \{k: \exists b < \infty \text{ such that } ||Gx||_{Tp} \leq k||x||_{Tp} + b, \ \forall T \geq 0, \ \forall x \in L^n_{pe}\}$$

If the set of k in (2) is empty, we set $\gamma_p(G) = \infty$. We define the *gain with zero bias* of G, denoted by $\zeta_p(G)$, by

3 $$\zeta_p(G) = \sup_{T \geq 0} \ \sup_{x_T \neq 0} \frac{||Gx||_{Tp}}{||x||_{Tp}}$$

If the indicated supremum does not exist, we set $\zeta_p(G) = \infty$. The *incremental gain* of G is denoted by $\eta_p(G)$, and is defined by

4 $$\eta_p(G) = \sup_{T \geq 0} \ \sup_{x_T \neq y_T} \frac{||Gx-Gy||_{Tp}}{||x-y||_{Tp}}$$

If the supremum in (4) does not exist, we set $\eta_p(G) = \infty$.

In the above definitions, we recognize that the constants $\gamma_p(G)$, $\zeta_p(G)$, and $\eta_p(G)$ depend not only on the operator G, but also on the value of p .

Note that, in general, $\gamma_p(G) \le \zeta_p(G)$, and $\zeta_p(G) \le \eta_p(G)$ whenever $G(0) = 0$. Also, if $\gamma_p(G)$ is finite, then G maps the <u>unextended</u> space L_p^n into the <u>unextended</u> space L_p^m . (However, the converse is not true; the operator $G:L_{\infty e} \to L_{\infty e}$ defined by $(Gx)(t) = x^2(t)$ maps L_∞ into L_∞ , but $\gamma_\infty(G) = \infty$). Finally, it is easy to show that if G is linear, then $\gamma_p(G) = \zeta_p(G) = \eta_p(G)$.

It is routine to verify that, given operators G_1 and G_2 defined on appropriate spaces, we have

5 $$\gamma_p(G_1 G_2) \le \gamma_p(G_1) \gamma_p(G_2)$$

6 $$\zeta_p(G_1 G_2) \le \zeta_p(G_1) \zeta_p(G_2)$$

7 $$\eta_p(G_1 G_2) \le \eta_p(G_1) \eta_p(G_2)$$

In the above definitions, we assume that G is an <u>operator</u> mapping L_{pe}^n into L_{pe}^m; i.e., we assume that, corresponding to <u>every</u> x in L_{pe}^n, there is a <u>unique</u> element Gx in L_{pe}^m. However, even if G is a <u>relation</u> (i.e., G is a subset of $L_{pe}^n \times L_{pe}^m$), one can still define $\gamma_p(G)$, $\zeta_p(G)$, and $\eta_p(G)$, by making only slight modifications to (2), (3), and (4).

Before proceeding to the actual computation of the constants $\gamma_p(G)$, $\zeta_p(G)$, and $\eta_p(G)$ for various types of operators, we present a useful general result.

8 <u>Lemma</u> Suppose $G:L_{pe}^n \to L_{pe}^m$ is causal. Then

9 $$\gamma_p(G) = \inf \{k: \exists b < \infty \text{ such that } \|Gx\|_p \le k\|x\|_p + b, \ \forall x \in L_p^n\}$$

10 $$\zeta_p(G) = \sup_{\substack{x \neq 0 \\ x \in L_p^n}} \frac{||Gx||_p}{||x||_p}$$

11 $$\eta_p(G) = \sup_{\substack{x \neq y \\ x,y \in L_p^n}} \frac{||Gx-Gy||_p}{||x-y||_p}$$

Proof We prove only (9); the proofs of (10) and (11) are entirely similar. To prove (9), it is enough to prove that G satisfies

12 $$||Gx||_p \leq k||x||_p + b, \quad \forall x \in L_p^n$$

for a particular choice of the constants k and b, if and only if it satisfies

13 $$||Gx||_{Tp} \leq k||x||_{Tp} + b, \quad \forall T \geq 0, \quad \forall x \in L_{pe}^n$$

for the same k and b. The "if" part follows readily by letting $T \to \infty$ in (13), and is true even if G is not causal. To prove the "only if" part, suppose G is causal and satisfies (12), and let $x \in L_{pe}^n$. Then we have

14 $$||Gx||_{Tp} = ||Gx_T||_{Tp} \leq ||Gx_T||_p \leq k||x_T||_p + b$$
$$= k||x||_{Tp} + b$$

This proves (9). □

The value of Lemma (8) is in showing that if $G:L_{pe}^n \to L_{pe}^m$ is causal, then the constants $\gamma_p(G)$, $\zeta_p(G)$ and $\eta_p(G)$ can be calculated by examining only the behaviour of G on the unextended space L_p^n.

3.1.2 Calculation of Gain for Linear Integral Operators

In this subsection, we show how to calculate $\zeta_p(G)$

for a given linear integral operator G. Note that, since G is linear, we have $\gamma_p(G) = \zeta_p(G) = \eta_p(G)$.

15 Lemma Consider an operator $G: L_{1e} \to L_{1e}$ defined by

16 $$(Gx)(t) = \int_0^t g(t,\tau)\, x(\tau)\, d\tau$$

where $g(.,.)$ is measurable and satisfies the condition

17 $$\sup_{\tau \in [0,T]} \int_\tau^T |g(t,\tau)|\, dt \triangleq C_T < \infty, \quad \forall T \geq 0$$

Then we have

18 $$\gamma_1(G) = \sup_{\tau \geq 0} \int_\tau^\infty |g(t,\tau)|\, dt$$

19 $$\gamma_\infty(G) = \sup_{t \geq 0} \int_0^t |g(t,\tau)|\, d\tau$$

20 $$\gamma_p(G) \leq [\gamma_1(G)]^{1/p}\, [\gamma_\infty(G)]^{1/q} \quad \forall p \in (1,\infty), \text{ where } q = p/(p-1)$$

Proof First of all, we show that the condition (17) is enough to guarantee that G does indeed map L_{1e} into L_{1e}. Accordingly, suppose $x \in L_{1e}^n$, so that

21 $$\int_0^T |x(\tau)|\, d\tau < \infty, \quad \forall T \geq 0$$

Then, from (17) and (21), we get

22 $$\int_0^T |(Gx)(t)|\, dt \leq \int_0^T \int_0^t |g(t,\tau)|\, |x(\tau)|\, d\tau\, dt$$

$$= \int_0^T |x(\tau)| \int_\tau^T |g(t,\tau)|\, dt\, d\tau$$

$$\leq C_T \int_0^T |x(\tau)| \, d\tau < \infty, \quad \forall T \geq 0$$

where the interchange of the order of integration is justified by Fubini's theorem [Dun. 1, p. 190]. Hence G maps L_{1e} into L_{1e} .

To prove (18), suppose first of all that

23 $$\sup_{\tau \geq 0} \int_\tau^\infty |g(t,\tau)| \, dt < \infty$$

Then, by letting $T = \infty$ in (22), it is clear that G maps L_1 into L_1, and that

24 $$\gamma_1(G) \leq \sup_{\tau \geq 0} \int_\tau^\infty |g(t,\tau)| \, dt < \infty$$

The proof that equality holds in (24) is facilitated by a bit of "hand-waving" using the unit impulse distribution $\delta(.)$. Once the basic idea of the proof is demonstrated, it will become clear that the argument can be made mathematically precise, by using a sequence of L_1-functions that have unit L_1-norm and converge to $\delta(.)$ in the sense of distributions. Suppose we let

25 $$x(t) = \delta(t-\tau_0)$$

for a fixed $\tau_0 \geq 0$. Then we have

26 $$(Gx)(t) = g(t,\tau_0)$$

Since $\delta(t-\tau_0)$ can be arbitrarily closely approximated, in the sense of distributions, by an L_1-function having unit L_1-norm, we conclude that

27 $$g(.,\tau_0) \in L_1$$

and moreover

28 $$\gamma_1(G) = \zeta_1(G) \geq ||g(.,\tau_0)||_1 = \int_0^\infty |g(t,\tau_0)|\, dt = \int_{\tau_0}^\infty |g(t,\tau_0)|\, dt$$

where in the last step we use the fact that $g(t,\tau) = 0$ whenever $t < \tau$. Since (28) holds for every $\tau_0 \geq 0$, we have

29 $$\gamma_1(G) \geq \sup_{\tau_0 \geq 0} \int_{\tau_0}^\infty |g(t,\tau_0)|\, dt$$

Now, (24) and (29) together prove (18).

Next, we prove (19). Suppose first of all that

30 $$\sup_t \int_0^t |g(t,\tau)|\, d\tau \triangleq d < \infty$$

and let $x \in L_\infty$. Then we have

31 $$|(Gx)(t)| \leq \int_0^t |g(t,\tau)|\ |x(\tau)|\, d\tau \leq d\ ||x(.)||_\infty$$

which shows that

32 $$\gamma_\infty(G) \leq \sup_t \int_0^t |g(t,\tau)|\, d\tau$$

To prove the opposite inequality, fix $t_0 \geq 0$, and let

33 $$x(\tau) = \text{sign } g(t_0,\tau) = \begin{cases} 1 & \text{if } g(t_0,\tau) \geq 0 \\ -1 & \text{if } g(t_0,\tau) < 0 \end{cases}$$

Then clearly $x(.) \in L_\infty$ and $||x(.)||_\infty = 1$. Thus

34 $$\gamma_\infty(G) \geq ||(Gx)(.)||_\infty = \sup_t \left| \int_0^t g(t,\tau)\, x(\tau)\, d\tau \right| \geq \int_0^{t_0} g(t_0,\tau)\, x(\tau)\, d\tau = \int_0^{t_0} |g(t_0,\tau)|\, d\tau$$

from the way in which $x(.)$ is chosen. Since (34) holds for every t_0, we have

35 $$\gamma_\infty(G) \geq \sup_{t_0} \int_0^{t_0} |g(t_0,\tau)|\, d\tau$$

Now, (19) follows from (32) and (35).

Finally, to prove (20), let $p \in (1,\infty)$, and let $x(.) \in L_p$. Then, letting $q = p/(p-1)$ and applying Hölder's inequality [Dun. 1, p. 119], we get

36 $$|(Gx)(t)| \leq \int_0^t |g(t,\tau)| \; |x(\tau)|\, d\tau$$

$$= \int_0^t |g(t,\tau)|^{1/q + 1/p} |x(\tau)|\, d\tau$$

$$\leq \{\int_0^t |g(t,\tau)|\, d\tau\}^{1/q} \{\int_0^t |g(t,\tau)| \; |x(\tau)|^p\, d\tau\}^{1/p}$$

37 $$|(Gx)(t)|^p \leq [\gamma_\infty(G)]^{p/q} \int_0^t |g(t,\tau)| \; |x(\tau)|^p\, d\tau$$

38 $$\int_0^\infty |(Gx)(t)|^p\, dt \leq [\gamma_\infty(G)]^{p/q} \int_0^\infty \int_0^t |g(t,\tau)| \; |x(\tau)|^p\, d\tau$$

$$\leq [\gamma_\infty(G)]^{p/q} \; \gamma_1(G) \int_0^\infty |x(\tau)|^p\, d\tau$$

$$= [\gamma_\infty(G)]^{p/q} \; \gamma_1(G) \; ||x(.)||_p^p$$

Raising both sides of (38) to the power $1/p$ gives

39 $$||(Gx)(.)||_p \leq [\gamma_\infty(G)]^{1/q} \; [\gamma_1(G)]^{1/p} \; ||x(.)||_p$$

which establishes (20). □

Lemma (15) shows that, for operators of the form (16), the gains $\gamma_1(G)$ and $\gamma_\infty(G)$ can be calculated exactly, whereas the gain $\gamma_p(G)$ can be estimated for all intermediate values of p. Note that as $p \to \infty$, the right side of (20) approaches $\gamma_\infty(G)$, while for $p \to 1$, it approaches $\gamma_1(G)$.

In proving (18) and (19), we made essential use of the fact that G maps <u>scalar</u> inputs into <u>scalar</u> outputs. If, on the other hand, G maps L^n_{1e} into L^m_{1e}, we can only obtain an upper bound for $\gamma_1(G)$, though it is still possible to calculate $\gamma_\infty(G)$.

40 <u>Lemma</u> Suppose $G : L^n_{1e} \to L^m_{1e}$ is of the form

41 $$(Gx)(t) = \int_0^t G(t,\tau)\, x(\tau)\, d\tau$$

where $G(.,.)$ is measurable and satisfies

42 $$\sup_{\tau \in [0,T]} \int_t^T ||G(t,\tau)||\, dt < \infty, \quad \forall T \geq 0$$

where $||.||$ denotes the matrix norm induced by the Euclidean norm. Then we have

43 $$\gamma_1(G) \leq \sup_{\tau \geq 0} \int_\tau^\infty ||G(t,\tau)||\, dt$$

44 $$\gamma_\infty(G) = \sup_{t \geq 0} \int_0^t ||G(t,\tau)||\, d\tau$$

45 $$\gamma_p(G) \leq [\gamma_1(G)]^{1/p}\, [\gamma_\infty(G)]^{1/q} \quad \forall\, p \in (1,\infty),$$

$$\text{where} \quad q = p/(p-1)$$

<u>Proof</u> The proofs of (43) and (45) are essentially the same as those of (24) and (20), respectively. To prove (44), we proceed just as in the proof of (19). First of all, as in (32), we have

46 $$\gamma_\infty(G) \leq \sup_{t \geq 0} \int_0^t ||G(t,\tau)|| \, d\tau$$

To prove the reverse inequality, let $t_0 \geq 0$ be arbitrary, let $e(\tau)$ be any unit vector such that $||G(t_0,\tau)\, e(\tau)|| = ||G(t_0,\tau)||$, and let

47 $$x(\tau) = e(\tau) \;, \quad \forall \tau$$

Then $x(.) \in L_\infty^n$ and $||x(.)||_\infty = 1$. Moreover,

48 $$\gamma_\infty(G) \geq ||(Gx)(.)||_\infty \geq ||(Gx)(t_0)|| = \int_0^\infty ||G(t_0,\tau)|| \, d\tau$$

Since (48) holds for every $t_0 \geq 0$, we have the reverse of (46), and (44) follows.

In the special case of linear <u>convolution</u> operators, one can obtain somewhat better estimates than those in Lemmas (15) and (40). One can also study a wider class of operators than those of the form (16). To do this, we recall Definition (2.1.31), where the set A is defined as the set of all distributions of the form

49 $$f(t) = \begin{cases} 0, & t < 0 \\ \sum_{i=0}^{\infty} f_i \, \delta(t-t_i) + f_a(t), & t \geq 0 \end{cases}$$

where $\delta(.)$ denotes the unit delta distribution, $0 \leq t_0 < t_1 \ldots$ are real constants, $\{f_i\} \in \ell_1$, and $f_a(.) \in L_1$; we also define

50 $$||f(.)||_A = \sum_{i=0}^{\infty} |f_i| + \int_0^\infty |f_a(t)| \, dt$$

51 <u>Lemma</u> Consider an operator $G : L_{1e} \to L_{1e}$ of the form

52 $$(Gx)(t) = \int_0^t g(t-\tau)\, x(\tau)\, d\tau = \sum_{i=0}^{\infty} g_i \, x(t-t_i) + \int_0^t g_a(t-\tau)\, x(\tau)\, d\tau$$

where $g(.) \in A$. Then

53 $$\gamma_1(G) = \gamma_\infty(G) = ||g(.)||_A$$

54 $$\gamma_p(G) \leq ||g(.)||_A \ , \quad \forall p \in [1,\infty]$$

Proof First of all, if $g_i = 0 \ \forall i$, i.e., if $g(.)$ actually belongs to L_1, then (53) and (54) follow directly from Lemma (15). The more general case considered here is proved in [Des. 2. p. 247]. □

We now present a sequence of three lemmas, whose purpose is to present an explicit expression for $\gamma_2(G)$.

55 Lemma Suppose $G : L_p(R) \to L_p(R)$ is a convolution operator defined by

56 $$(Gx)(t) = \int_{-\infty}^{t} g(t-\tau)\ x(\tau)\ d\tau$$

where $g(.) \in A$, and suppose $p < \infty$. Then

57 $$\gamma = \gamma_+$$

where

58 $$\gamma = \sup_{x \in L_p(R)/\{0\}} \frac{||\ Gx\ ||_p}{||x||_p}$$

59 $$\gamma_+ = \sup_{x \in L_p(R_+)/\{0\}} \frac{||\ Gx\ ||_p}{||x||_p}$$

Remarks The purpose of this lemma is to show that the gain of G viewed as an operator from $L_p(R)$ into $L_p(R)$ is exactly the same as its gain when viewed as an operator from $L_p(R_+)$ into $L_p(R_+)$.

Proof Clearly $\gamma \geq \gamma_+$. To prove the opposite, let $\varepsilon > 0$ be given, and select $\delta \in (0,1)$ such that

60 $$(1 - \tfrac{\varepsilon}{2} - \delta^{1/p})/(1-\delta)^{1/p} \geq 1 - \varepsilon$$

This can always be done, because as $\delta \to 0$, the left side of (60) approaches $1-(\varepsilon/2) > 1-\varepsilon$. Next, select $x \in L_p(R)$ such that $||x||_p = 1$ and $||Gx||_p \geq [1-(\varepsilon/2)]\gamma$; then select $T < \infty$ such that

61 $$\int_{-\infty}^{T} |x(t)|^p \, dt = \delta$$

This can always be done, because the left side of (61) is a continuous nondecreasing function of T that approaches 0 as $T \to -\infty$, and 1 as $T \to \infty$. Now let $x(t) = x_1(t) + x_2(t)$, where $x_1(t)$ vanishes for $t < T$ and $x_2(t)$ vanishes for $t > T$. Then clearly

62 $$||x||_p^p = ||x_1||_p^p + ||x_2||_p^p$$

so that

63 $$||x_1||_p^p = 1-\delta \;, \; ||x_2||_p^p = \delta$$

Next, we have

64 $$||Gx_1||_p = ||Gx - Gx_2||_p \geq ||Gx||_p - ||Gx_2||_p$$
$$\geq (1 - \tfrac{\varepsilon}{2})\,\gamma - \delta^{1/p}\,\gamma = (1 - \tfrac{\varepsilon}{2} - \delta^{1/p})\,\gamma$$

65 $$\frac{||Gx_1||_p}{||x_1||_p} \geq \frac{1 - \frac{\varepsilon}{2} - \delta^{1/p}}{(1-\delta)^{1/p}}\,\gamma \geq (1-\varepsilon)\gamma$$

Now, observe that x_1, Gx_1 actually belong to $L_2[T,\infty)$.

Define

66 $$y(t) = x_1(t-T) \;, \quad \forall t \in R_+$$

Then $y \in L_p(R_+)$, and $||y||_p = ||x_1||_p$. Also, since G is a time-invariant operator,

67 $$(Gy)(t) = (Gx_1)(t-T) \;, \quad \forall t \in R_+$$

so that $Gy \in L_p(R_+)$, and

68 $$\frac{||Gy||_p}{||y||_p} \geq (1-\varepsilon)\gamma$$

In effect, what we have shown is that, given any $\varepsilon > 0$, we can construct a function $y \in L_p(R_+)$ such that (68) holds. This proves that $\gamma_+ = \gamma$. □

69 Lemma Suppose $G : L_2(R) \to L_2(R)$ is defined by (56), where $g(.) \in A$. Then

70 $$\sup_{x \in L_2(R)/\{0\}} \frac{||Gx||_2}{||x||_2} = \sup_\omega |\hat{g}(j\omega)|$$

Proof For the sake of convenience, define

71 $$c_2 \triangleq \sup_\omega |\hat{g}(j\omega)|$$

Given any $\varepsilon > 0$, select $\omega_o \in R$ such that $|\hat{g}(j\omega_o)| \geq c_2 - (\varepsilon/2)$. Since $\hat{g}(-j\omega_o)$ is the complex conjugate of $\hat{g}(j\omega_o)$, we also have $|\hat{g}(-j\omega_o)| \geq c_2 - (\varepsilon/2)$. Now, let

72 $$x_\sigma(t) = \exp(-t^2/4\sigma^2) \cos(\omega_o t)/(2\pi\sigma^2)^{1/4}$$

where $\sigma > 0$. Then $x_\sigma(.) \in L_2(R)$, $||x_\sigma(.)||_2 = 1$, and

73 $$\hat{x}_\sigma(j\omega) = [\hat{\phi}(j\omega-j\omega_o) + \hat{\phi}(j\omega+j\omega_o)]/2$$

where

74 $$\hat{\phi}(j\omega) = (8\pi\sigma^2)^{1/4} e^{-\sigma^2\omega^2}$$

Also

75 $$(Gx_\sigma)(j\omega) = \hat{g}(j\omega)\,\hat{x}_\sigma(j\omega)$$

Now, by letting $\sigma \to \infty$, it is clear that one can essentially concentrate the frequency spectrum $|\hat{x}_\sigma(j\omega)|$ around the two frequencies $\omega = \pm\omega_o$. By Parseval's equality, we have

76 $$||Gx_\sigma||_2^2 = \frac{1}{2\pi}\int_{-\infty}^{\infty} |\hat{g}(j\omega)|^2 \, |\hat{x}_\sigma(j\omega)|^2 \, d\omega$$

Hence, as $\sigma \to \infty$, $||Gx_\sigma||_2 \to [|\hat{g}(j\omega_o)| + |\hat{g}(-j\omega_o)|]/2 \geq c_2-(\varepsilon/2)$. So we can select a finite σ such that $||Gx_\sigma|| \geq c_2-\varepsilon$. Since this can be done for any ε, we see that

77 $$\sup_{x \in L_2(R)/\{0\}} \frac{||Gx||_2}{||x||_2} \geq c_2$$

To prove the opposite inequality, we gain use Parseval's equality. This gives, for all $x \in L_2(R)$

78 $$\begin{aligned} ||Gx||_2^2 &= \frac{1}{2\pi} \int_{-\infty}^{\infty} |Gx(j\omega)|^2 \, d\omega \\ &= \frac{1}{2\pi} \int_{-\infty}^{\infty} |\hat{g}(j\omega) \, \hat{x}(j\omega)|^2 \, d\omega \\ &\leq \sup_\omega |\hat{g}(j\omega)|^2 \, \frac{1}{2\pi} \int_{-\infty}^{\infty} |\hat{x}(j\omega)|^2 \, d\omega \end{aligned}$$

which shows that

79 $$\sup_{x \in L_2(R)/\{0\}} \frac{||Gx||_2}{||x||_2} \leq \sup_\omega |\hat{g}(j\omega)| = c_2$$

Now, (77) and (79) together establish (70). □

80 Lemma Consider an operator $G : L_2 \to L_2$ of the form

81 $$(Gx)(t) = \int_0^t g(t-\tau) \, x(\tau) \, d\tau$$

where $g(.) \in A$. Then

82 $$\gamma_2(G) = \sup_\omega |\hat{g}(j\omega)|$$

Proof Clearly, G defined by (56) is an extension to $L_2(R)$ of G defined by (81). The equality (82) now follows readily from Lemmas (69) and (55). □

We close out the discussion of the gains of linear integral operators by presenting a result on multi-input-multi-output convolution operators.

83 Lemma Suppose $G : L_{1e}^n \to L_{1e}^m$ is of the form

84 $$(Gx)(t) = \int_0^t G(t-\tau)\, x(\tau)\, d\tau$$

where $G(.) \in L_1^{m\times n}$. Then

85 $$\gamma_1(G) \leq \gamma_\infty(G) = \int_0^\infty ||G(t)||\, dt$$

86 $$\gamma_2(G) = \sup_\omega \lambda_{max} [\hat{G}^+(j\omega)\hat{G}(j\omega)]^{1/2} \triangleq c_2$$

where '+' denotes the conjugate transpose, and λ_{max} denotes the largest eigenvalue of a matrix.

Proof The condition (85) follows routinely from (43) and (44). To prove (86), observe that, for any $x \in L_2^n$, we have

87 $$||Gx||_2^2 = \frac{1}{2\pi} \int_{-\infty}^\infty [\hat{G}\hat{x}(j\omega)]^+ [\hat{G}\hat{x}(j\omega)]\, d\omega$$
$$= \frac{1}{2\omega} \int_{-\infty}^\infty \hat{x}^+(j\omega)\hat{G}^+(j\omega)\hat{G}(j\omega)\hat{x}(j\omega)\, d\omega$$
$$\leq \frac{1}{2\pi} c_2^2 \int_{-\infty}^\infty \hat{x}^+(j\omega)\hat{x}(j\omega)\, d\omega$$
$$= c_2^2\, ||x||_2^2$$

which shows that $\gamma_2(G) \leq c_2$. The proof that $\gamma_2(G) = c_2$ now follows along the steps of Lemmas (55) and (69). □

Remark Note that (86) holds even if $G(.) \in A^{m\times n}$.

3.1.3. Calculation of Gain for Memoryless Nonlinearities

In this subsection, we study operators $G : L_{1e}^n \to L_{1e}^m$ of the form

88 $$(Gx)(t) = G(t,x(t))$$

where $G(.,.): R_+ x R^n \to R^m$ is continuous.

89 <u>Lemma</u> Suppose there exists a constant k such that

90 $$||G(t,v)|| \leq k \, ||v||, \quad \forall v \in R^n, \quad \forall t \geq 0$$

where $||.||$ in (90) denotes the Euclidean norm. Then

91 $$\zeta_p(G) < k, \quad \forall p \in [1,\infty]$$

<u>Proof</u> For any $p \in [1,\infty)$, we have

92 $$\int_0^T ||G(t,x(t))||^p \, dt \leq k^p \int_0^T ||x(t)||^p \, dt \;, \quad \forall T \geq 0$$

so that (91) follows readily by raising both sides of (92) to the power $1/p$. Next, for $p = \infty$,

93 $$\underset{t \in [0,T]}{\text{ess. sup}} \; ||G(t,x(t))|| \leq k \underset{t \in [0,T]}{\text{ess.sup}} \; ||x(t)||$$

so that $\zeta_\infty(G) \leq k$. □

94 <u>Lemma</u> Suppose there exists a constant α such that

95 $$||G(t,u)-G(t,v)|| \leq \alpha \, ||u-v||, \quad \forall u,v \in R^n \;, \quad \forall t \geq 0$$

Then

96 $$\eta_p(G) \leq \alpha \;, \quad \forall p \in [1,\infty]$$

<u>Proof</u> Similar to that of Lemma (89). □

Next, we present a specialized result for the case $p = \infty$.

97 <u>Lemma</u> Suppose there exist constants c and b such that

98 $$||G(t,v)|| \leq c\,||v|| + b\ , \quad \forall v \in R^n\ , \quad \forall t \geq 0$$

Then

99 $$\gamma_\infty(G) \leq c$$

<u>Proof</u> Clearly, if $x \in L_{\infty e}$, then

100 $$\underset{t \in [0,T]}{\text{ess.sup}} ||G(t,x(t))|| \leq c \underset{t \in [0,T]}{\text{ess.sup}} ||x(t)|| + b$$

which proves (99). □

Notice that (98) does <u>not</u> imply that $\gamma_p(G) \leq c$, for $p \neq \infty$.

101 <u>Definition</u> A function $G : R_+ \times R^n \to R^n$ is said to <u>belong to the sector $[\alpha,\beta]$</u> if

102 $$[G(t,v)-\alpha v]'\ [G(t,v)-\beta v] \leq 0 \quad \forall v \in R^n\ , \quad \forall t \geq 0$$

<u>Remarks</u> In the case $n = 1$, G belongs to the sector $[\alpha,\beta]$ if, for each $t \geq 0$, the graph of $G(t,.)$ lies in the shaded region shown in Figure 3.1.

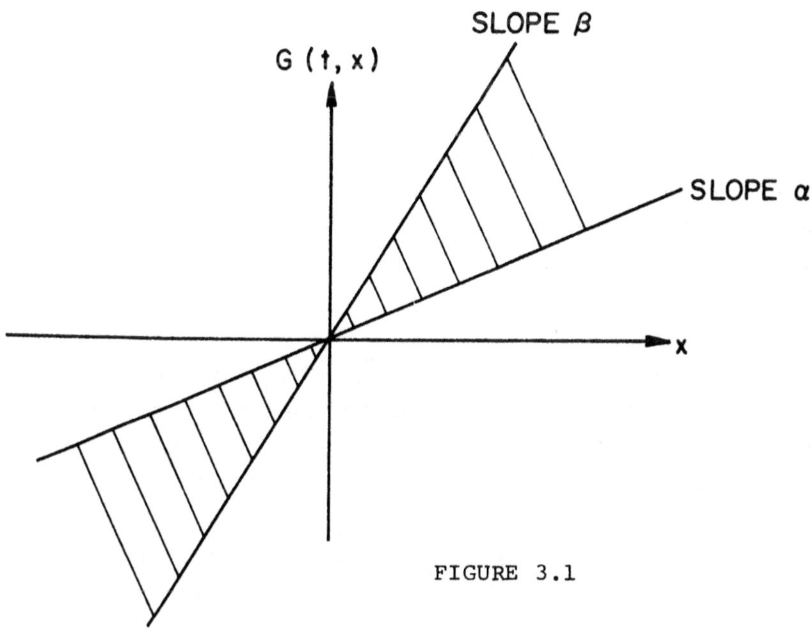

FIGURE 3.1

103 **Lemma** Suppose $G : R_+ \times R^n \to R^n$ belongs to the sector $[\alpha,\beta]$, and let G denote the corresponding operator defined by (88). Then

104 $$\gamma_2(G-cI) \le r \quad , \quad \forall p \in [1,\infty]$$

where I denotes the identity operator on L^n_{2e} , and

105 $$c = [\beta+\alpha]/2 \quad , \quad r = [\beta-\alpha]/2$$

Proof By "completing the square," (102) can be equivalently expressed as

106 $$||G(t,v)-cv||^2 \le r^2 \, v'v \quad \forall v \in R^n \ , \quad \forall t \ge 0$$

where $||.||$ denotes the Euclidean norm. The bound (104) now readily follows from (106). □

3.2 DISSIPATIVITY AND PASSIVITY

In the previous section, we have introduced the concept of gain, which can be applied to operators $G : L^n_{pe} \to L^m_{pe}$ for all values of $p \in [1,\infty]$. In this section, we define the concepts of passivity and dissipativity, which can only be applied to operators $G : L^n_{2e} \to L^m_{2e}$. In other words, it is essential to have a truncated inner product structure on the input and output spaces. To keep the exposition simple, we assume throughout that $m = n$, i.e. that $G : L^n_{2e} \to L^n_{2e}$. However, the modifications required to handle the case $m \neq n$ are readily apparent.

1 **Definition** Let $G : L^n_{2e} \to L^n_{2e}$, and let $Q,R,S \in R^{n\times n}$ with Q and R symmetric. We say that the operator G is dissipative with respect to the triplet (Q,R,S), or simply (Q,R,S)-dissipative if

2 $$\langle Gx,QGx\rangle_T + \langle x,Rx\rangle_T + \langle Gx,Sx\rangle_T \ge 0 \ , \ \forall T \ge 0 \ , \ \forall x \in L^n_{2e}$$

Before discussing the implications of dissipativity, we first state a general result for causal operators, analogous to Lemma (3.1.8).

3 Lemma Suppose $G : L_{2e}^n \to L_{2e}^n$ is causal, and that Q is negative semidefinite. Then G is (Q,R,S)-dissipative if and only if

4 $$\langle Gx,QGx\rangle + \langle x,Rx\rangle + \langle Gx,Sx\rangle \geq 0 \ , \quad \forall x \in L_2^n$$

Proof First of all, suppose (2) holds (i.e., G is (Q,R,S)-dissipative), and let $x \in L_2^n$. Then (4) follows by letting $T \to \infty$.

Conversely, suppose G satisfies (4), and let $x \in L_{2e}^n$. Then $x_T \in L_2$ for all T , and by (4) we have

5 $$\langle Gx_T,QGx_T\rangle + \langle x_T,Rx_T\rangle + \langle Gx_T,Sx_T\rangle \geq 0$$

However, since Q is negative semidefinite and G is causal, we have

6 $$\langle Gx_T,QGx_T\rangle \leq \langle Gx_T,QGx_T\rangle_T = \langle Gx,QGx\rangle_T$$

Similarly, by the causality of G , we have

7 $$\langle Gx_T,Sx_T\rangle = \langle Gx,Sx\rangle_T$$

Clearly,

8 $$\langle x_T,Rx_T\rangle = \langle x,Rx\rangle_T$$

Substituting into (5) gives

9 $$\langle Gx,QGx\rangle_T + \langle x,Rx\rangle_T + \langle Gx,Sx\rangle_T \geq 0$$

which establishes (2). □

As with Lemma (3.1.8), the significance of Lemma (3) lies in the fact that, for a causal operator G , one can determine whether or not G is (Q,R,S)-dissipative (where Q is negative semidefinite) by examining only the behaviour of G on the unextended space L_2^n.

We now turn to the implications of dissipativity. Note that, in the interests of brevity, we use $||\cdot||$ to denote $||\cdot||_2$ throughout this section.

10 **Lemma** Suppose $G : L_2^n \to L_2^n$ is (Q,R,S) dissipative where Q is negative definite. Then $\zeta_2(G) < \infty$.

Proof Let $-\lambda$ denote the largest (least negative) eigenvalue of Q. Then (2) becomes

11 $$\lambda||Gx||_T^2 - \beta||x||_T^2 \leq \alpha||x||_T\ ||Gx||_T \quad \forall T \geq 0 \quad \forall x \in L_{2e}^n$$

where $\beta = ||R||$, $\alpha = ||S||$. By completing the square, (11) can be manipulated into the form

12 $$||Gx||_T \leq \zeta\ ||x||_T, \quad \forall T \geq 0, \quad \forall x \in L_{2e}^n$$

which shows that $\zeta_2(G) < \infty$. □

13 **Lemma** Suppose $\zeta_2(G) < \infty$. Then G is dissipative with respect to the triplet $(-I_n, \zeta_2^2(G) I_n, 0_n)$, where 0_n denotes the null matrix of dimension $n\times n$ and I_n denotes the identity matrix of dimension $n\times n$.

Proof Obvious on noting that

14 $$||Gx||_T^2 \leq [\zeta_2(G)]^2\ ||x||_T^2 \quad \forall T \geq 0, \quad \forall x \in L_{2e}^n \quad □$$

15 **Lemma** Suppose $G : L_{2e}^n \to L_{2e}^n$ is dissipative with respect to (Q,R,S), and the $\zeta_2(G) < \infty$. Then it is also dissipative with respect to $(Q-\alpha I_n, R+\alpha\zeta_2^2(G) I_n, S)$, for all $\alpha \geq 0$.

Proof Obvious, by combining inequalities (2) and (14).□

The next definition introduces an important concept which was historically the first type of dissipativity to be studied in detail.

16 <u>Definition</u> An operator $G : L_{2e}^n \to L_{2e}^n$ is said to be <u>passive</u> if it is dissipative with respect to $(0_n, 0_n, I_n)$, i.e. if

17 $$\langle x, Gx\rangle_T \geq 0 \ , \quad \forall T \geq 0 \ , \quad \forall x \in L_{2e}^n$$

G is said to be <u>strictly passive</u> if it is dissipative with respect to $(0_n, -\varepsilon I_n, I_n)$ for some $\varepsilon > 0$, i.e., if

18 $$\langle x, Gx\rangle_T \geq \varepsilon \ ||x||_T^2 \ , \quad \forall T \geq 0 \ , \quad \forall x \in L_{2e}^n$$

We now consider how these various concepts apply to linear convolution operators and memoryless nonlinear operators. Due to shortage of space, we do not discuss the dissipativity of integral operators of the type (3.1.16). Instead, the reader is referred to [Wil. 4] for a discussion of the passivity of operators of the type (3.1.16).

19 <u>Lemma</u> Consider a convolution operator $G : L_{2e}^n \to L_{2e}^n$ of the form

20 $$Gx(t) = \int_0^t G(t-\tau) \ x(\tau) \ d\tau$$

where $G(.) \in A^{n \times n}$. Then G is dissipative with respect to $(0_n, -\delta I_n, I_n)$, where

21 $$\delta = \inf_{\omega \in R} \lambda_{min} \ [\hat{G}^+(j\omega) + \hat{G}(j\omega)]/2$$

<u>Proof</u> Clearly G is causal. Hence we can apply Lemma (3). Let $x \in L_2^n$; then by Parseval's theorem,

22 $$\langle x, Gx\rangle = \frac{1}{2\pi} \ \mathrm{Re} \int_{-\infty}^{\infty} \hat{x}^+(j\omega) \ \hat{G}(j\omega) \ \hat{x}(j\omega) \ d\omega$$

$$= \frac{1}{2\pi} \int_{-\infty}^{\infty} \hat{x}^+(j\omega) \ \frac{[\hat{G}^+(j\omega) + \hat{G}(j\omega)]}{2} \ \hat{x}(j\omega) \ d\omega$$

$$\geq \frac{\delta}{2\pi} \int_{-\infty}^{\infty} |\hat{x}(j\omega)|^2 \ d\omega = \delta \ ||x||^2$$

Hence (4) holds with $Q = 0_n$, $R = -\delta I_n$, $S = I_n$. By Lemma (3), G is $(0_n, -\delta I_n, I_n)$-dissipative. □

23 <u>Lemma</u> Suppose $G : L_{2e}^n \to L_{2e}^n$ is an operator of the form (3.1.85), where G belongs to the sector $[\alpha,\beta]$, i.e., $G(.,.)$ satisfies (3.1.99). Then G is dissipative with respect to $(-I_n, -\alpha\beta I_n, (\alpha+\beta) I_n)$.

<u>Proof</u> From (3.1.99), one can easily show that

24 $$\langle Gx-\alpha x, Gx-\beta x\rangle_T \leq 0$$

which can be readily manipulated to yield the desired result. □

3.3 CONDITIONAL GAIN AND CONDITIONAL DISSIPATIVITY

In this section, we shall be concerned with "unstable" operators, i.e., operators $G : L_{pe}^n \to L_{pe}^n$ which do <u>not</u> map L_p^n into L_p^m. We first introduce the basic definitions, and then discuss how they pertain to linear convolution operators.

1 <u>Definition</u> Suppose $G : L_{pe}^n \to L_{pe}^m$. Then we define

2 $$M_p(G) = \{x \in L_p^n : Gx \in L_p^n\}$$

The <u>conditional gain</u> of G, denoted by $\zeta_{cp}(G)$, is defined by*

3 $$\zeta_{cp}(G) = \sup_{x \in M_p(G)/\{0\}} \frac{\|Gx\|_p}{\|x\|_p}$$

If $n = m$, $p = 2$, the operator G is said to be <u>conditionally (Q,R,S)-dissipative</u> with respect to a given triplet (Q,R,S) if

4 $$\langle Gx,QGx\rangle + \langle x,Rx\rangle + \langle Gx,Sx\rangle \geq 0, \quad \forall x \in M_2(G)$$

---

*Note that $\zeta_{cp}(G)$ should properly be called "conditional gain with zero bias." However, this is too cumbersome, and we never use $\gamma_{cp}(G)$.

The following simple lemmas bring out some relationships between the concepts introduced in Definition (1) and the "unconditional" concepts introduced earlier.

5 <u>Lemma</u> Suppose $\zeta_p(G) < \infty$; then $M_p(G) = L_p^n$ and $\zeta_{cp}(G) = \zeta_p(G)$.

6 <u>Lemma</u> Suppose G is causal and that G is conditionally dissipative with respect to (Q,R,S) where $Q \leq 0$ (i.e., Q is negative semidefinite); suppose $M_2(G) = L_2^n$. Then G is (Q,R,S)-dissipative.

<u>Proof</u> The hypotheses imply that

7 $$\langle Gx, QGx\rangle + \langle x, Rx\rangle + \langle Gx, Sx\rangle \geq 0 \quad \forall x \in L_2^n$$

Since G is causal, it follows by Lemma (3.2.3) that G is (Q,R,S)-dissipative.□

8 <u>Lemma</u> Suppose G is causal, that G is conditionally (Q,R,S)-dissipative but not (Q,R,S)-dissipative, and suppose $Q \leq 0$. Then $M_2(G) \neq L_2^n$.

<u>Proof</u> This is just the contrapositive of Lemma (6).□

9 <u>Lemma</u> Suppose G is causal, that G is conditionally (Q,R,S)-dissipative where $Q < 0$ (i.e., Q is negative definite); suppose $M_2(G) \neq L_2^n$. Then G is not (Q,R,S)-dissipative.

<u>Proof</u> If G is (Q,R,S)-dissipative, then by Lemma (3.2.10), we have $\zeta_2(G) < \infty$, which shows that $M_2(G) = L_2^n$, by Lemma (5). Since $M_2(G) \neq L_2^n$, G cannot be (Q,R,S)-dissipative.□

We next introduce a class of operators that are unstable in a particular way. Such operators play an important role in the instability theorems of Chapters 8 and 9.

10 <u>Definition</u> An operator $G : L_{2e}^n \to L_{2e}^n$ is said to <u>belong to class U</u> if

(i) G is linear

(ii) There is an unbounded family of constants α_T such that

11 $$||Gx||_{T2} \leq \alpha_T \, ||x||_{T2} \, , \quad \forall x \in L_{2e}^n$$

(iii) $M_2(G) \neq L_2^n$

(iv) $\zeta_{c2}(G) < \infty$

12 <u>Lemma</u> Suppose $G : L_{2e}^n \to L_{2e}^n$ belongs to class U. Then $M_2(G)$ is a proper closed subspace of L_2^n .

<u>Proof</u> Since G is linear, $M_2(G)$ is a subspace of L_2^n . It is also a proper subspace of L_2^n since $M_2(G) \neq L_2^n$. All that remains to be shown is that it is a closed subspace. To do this, suppose $\{x_i\}_1^\infty$ is a sequence in $M_2(G)$ converging to $x^* \in L_2^n$; we must show that $x^* \in M_2(G)$. Let $y_i = Gx_i \in L_2^n$, $y^* = Gx^* \in L_{2e}^n$. Since $\{x_i\}$ is a Cauchy sequence, and since $\zeta_{c2}(G) < \infty$, it can be seen that $\{y_i\}$ is also a Cauchy sequence. Let $z \in L_2^n$ denote the limit of $\{y_i\}$. Now, from (11), we see that for each $T < \infty$, the sequence $\{(Gx_i)_T\}$ converges to y_T^* . On the other hand, since $\{y_i\}$ converges to z , we see that $\{y_{iT}\} = \{(Gx_i)_T\}$ converges to z_T . This shows that $y_T^* = z_T$ for all T , i.e., $y^* = z \in L_2^n$. Since $y^* = Gx^* \in L_2^n$ and $x^* \in L_2^n$, it follows that $x^* \in M_2(G)$. □

13 <u>Lemma</u> Suppose $G : L_{2e}^n \to L_{2c}^n$ belongs to class U. Then $M^\perp(G)$, defined by

14 $$M^\perp(G) = \{z \in L_2^n : \langle z,x\rangle = 0 \text{ for all } x \in M_2(G)\}$$

contains some nonzero elements.

<u>Proof</u> $M^\perp(G)$ is the orthogonal complement of $M_2(G)$, which is a proper closed subspace of L_2^n . □

The rest of this section is devoted to a characterization of the sets $M_p(G)$ and $M^\perp(G)$ in the case where G is a linear convolution operator. To do this, we need the concept of a right-coprime factorization.

15 <u>Definition</u> Given a Laplace transformable distribution $G(.)$ of dimension $n\times n$, the ordered pair $(\hat{N},\hat{D})$ is said to be a <u>right-coprime factorization</u> (r.c.f.) of $\hat{G}(.)$ if

(i) $\hat{N}, \hat{D} \in \hat{A}^{n\times n}$

(ii) $\hat{N}$ and $\hat{D}$ are <u>right-coprime</u>, i.e., there exist $\hat{P}, \hat{Q}$ in $\hat{A}^{n\times n}$ such that

16 $$\hat{P}(s)\,\hat{N}(s) + \hat{Q}(s)\,\hat{D}(s) = I\ , \quad \forall s \in C_+$$

17 $$\text{(iii)} \quad \hat{G}(s) = \hat{N}(s)\,[\hat{D}(s)]^{-1}\ , \quad \forall s \in C_+$$

Right-coprime factorizations have several useful properties, some of which are brought out below. Others can be found in [Vid. 1] and [Vid. 5].

18 <u>Lemma</u> If $\hat{G}(s)$ is the sum of an $n\times n$ matrix of rational functions of s and an element of $\hat{A}^{n\times n}$, then $\hat{G}(.)$ has an r.c.f.

For a proof of this result, see [Vid. 5]. Lemma (18) shows that the most commonly encountered type of "transfer function matrices" have r.c.f.'s.

19 <u>Lemma</u> Let $(\hat{N}_1,\hat{D}_1)$, $(\hat{N}_2,\hat{D}_2)$ be two different r.c.f.'s of $\hat{G}(.)$. Then there is an $\hat{R}(.) \in \hat{A}^{n\times n}$ such that $\hat{R}^{-1} \in \hat{A}^{n\times n}$, $\hat{N}_2 = \hat{N}_1\hat{R}$, $\hat{D}_2 = \hat{D}_1\hat{R}$.

<u>Remarks</u> Lemma (19) shows that an r.c.f. of $\hat{G}$ is unique to within a "regular" element $\hat{R}$ of the ring $\hat{A}^{n\times n}$, i.e., to within an element $\hat{R}$ in $\hat{A}^{n\times n}$ whose inverse also belongs to $\hat{A}^{n\times n}$.

<u>Proof</u> If $(\hat{N}_1,\hat{D}_1)$, $(\hat{N}_2,\hat{D}_2)$ are both r.c.f.'s of $\hat{A}^{n\times n}$, then by assumption, there exist $\hat{P}_1,\hat{Q}_1,\hat{P}_2,\hat{Q}_2$ in $\hat{A}^{n\times n}$ such that

20 $$\hat{P}_1\hat{N}_1 + \hat{Q}_1\hat{D}_1 = I$$

21 $$\hat{P}_2\hat{N}_2 + \hat{Q}_2\hat{D}_2 = I$$

Multiplying both sides of (20) by $\hat{D}_1^{-1}$ gives

22 $$\hat{P}_1\hat{N}_1\hat{D}_1^{-1} + \hat{Q}_1 = \hat{D}_1^{-1}$$

By assumption, $\hat{G} = \hat{N}_1\hat{D}_1^{-1} = \hat{N}_2\hat{D}_2^{-1}$. Hence

23 $$\hat{P}_1\hat{N}_2\hat{D}_2^{-1} + \hat{Q}_1 = \hat{D}_1^{-1}$$

Multiplying both sides of (23) by $\hat{D}_2$ gives

24 $$\hat{D}_1^{-1}\hat{D}_2 = \hat{P}_1\hat{N}_2 + \hat{Q}_1\hat{D}_2 \triangleq \hat{R} \in \hat{A}^{n\times n} ,$$

Now, (24) shows that $\hat{D}_2 = \hat{D}_1\hat{R}$, which implies that $\hat{N}_2 = \hat{N}_1\hat{R}$, since $\hat{N}_1\hat{D}_1^{-1} = \hat{N}_2\hat{D}_2^{-1}$. Also, (24) shows that $\hat{R} \in \hat{A}^{n\times n}$. Finally, an entirely analogous argument starting from (21) instead of (20) shows that $\hat{R}^{-1} \in \hat{A}^{n\times n}$. □

25 Lemma Suppose $(\hat{N},\hat{D})$ is an r.c.f. of $\hat{G}$, and suppose $\hat{H} \in \hat{A}^{n\times n}$. Then $(\hat{N} + \hat{H}\hat{D}, \hat{D})$ is an r.c.f. of $\hat{G} + \hat{H}$.

Proof By assumption, there exist $\hat{P},\hat{Q}$ in $\hat{A}^{n\times n}$ such that

26 $$\hat{P}\hat{N} + \hat{Q}\hat{D} = I$$

Hence

27 $$\hat{P}(\hat{N} + \hat{H}\hat{D}) + (\hat{Q} - \hat{P}\hat{H})\hat{D} = I$$

which shows that $\hat{N} + \hat{H}\hat{D},\hat{D}$ are right coprime. Finally, it is evident that $\hat{G} + \hat{H} = (\hat{N} + \hat{H}\hat{D}).\hat{D}^{-1}$. □

Lemma (25) states that if we are given a transfer function matrix $\hat{G}_1$ which can be expressed as the sum of a "stable" part $\hat{H}$ and an "unstable" part $\hat{G}$, then it is only necessary to find an r.c.f. of the unstable part $\hat{G}$, because, once we have an r.c.f. $(\hat{N},\hat{D})$ for $\hat{G}$, we readily have an r.c.f. for $\hat{G} + \hat{H}$, namely $(\hat{N} + \hat{H}\hat{D},\hat{D})$.

28 **Lemma** A point $s_o \in C_+$ is a singularity of $\hat{G}(.)$ if and only if det $\hat{D}(s_o) = 0$, where $(\hat{N},\hat{D})$ is any r.c.f. of $\hat{G}$.

Proof "if" Suppose det $\hat{D}(s_o) = 0$. Then some element of $[\hat{D}(s)]^{-1}$ becomes unbounded as $s \to s_o$ because otherwise det $[\hat{D}(s)]^{-1} = 1/\det \hat{D}(s)$ remains bounded as $s \to s_o$. Now select $\hat{P},\hat{Q}$ in $\hat{A}^{n\times n}$ such that

29 $$\hat{P}(s)\ \hat{N}(s) + \hat{Q}(s)\ \hat{D}(s) = I \quad , \quad \forall s \in C_+$$

Multiplying both sides by $[\hat{D}(s)]^{-1}$ gives

30 $$\hat{P}(s)\ \hat{G}(s) + \hat{Q}(s) = [\hat{D}(s)]^{-1}$$

As $s \to s_o$, the right side of (30) becomes unbounded, whence so must the left side. Since $\hat{P},\hat{Q} \in \hat{A}^{n\times n}$, $\hat{P}(s)$ and $\hat{Q}(s)$ remain bounded as $s \to s_o \in C_+$. Hence $\hat{G}(s)$ must become unbounded as $s \to s_o$, which shows that s_o is a singularity of $\hat{G}(.)$.

"only if" since $\hat{G} = \hat{N}\ \hat{D}^{-1}$, we can write

31 $$\hat{G}(s) = \hat{N}(s).\ \text{Adj}\ \hat{D}(s)\ /\ \det \hat{D}(s)$$

where Adj $\hat{D}(s)$ denotes the matrix of cofactors of $\hat{D}(s)$. Since each element of Adj $\hat{D}(s)$ is obtained by multiplying or adding elements of $\hat{A}$, it follows that Adj $\hat{D}(s) \in \hat{A}^{n\times n}$, whence $\hat{N}(s)$. Adj $\hat{D}(s) \in \hat{A}^{n\times n}$ and is bounded over C_+. So, if $\hat{G}(s)$ becomes unbounded as $s \to s_o$, we must have det $\hat{D}(s) \to 0$ as $s \to s_o$, i.e., det $\hat{D}(s_o) = 0$. □

With this background on r.c.f.'s , we can characterize the set $M_p(G)$ whenever G is a convolution operator with "transfer function matrix" $\hat{G}(.)$, which has an r.c.f. $(\hat{N},\hat{D})$.

32 **Theorem** Suppose $G : L^n_{pe} \to L^n_{pe}$ is a convolution operator of the form

33 $$(Gx)(t) = \int_0^t G(t-\tau)\ x(\tau)\ d\tau = (G*x)(t)$$

where $G(.)$ is a Laplace transformable distribution, and $\hat{G}(.)$ has the r.c.f. $(\hat{N},\hat{D})$. Then, for every $p \in [1,\infty]$,

34 $$M_p(G) = \{x : x = D*z \quad \text{for some} \quad z \in L_p^n\}$$

$$= \{x : x(t) = \int_0^t D(t-\tau)\, z(\tau)\, d\tau \text{ for some } z(.) \in L_p^n\}$$

Proof Let $D : L_{pe}^n \to L_{pe}^n$ be the operator mapping z into $D*z$. Then the right side of (34) is simply $D(L_p^n)$, the image of L_p^n under the operator D . We claim that $M_p(G) = D(L_p^n)$. To prove this claim, observe first of all that $M_p(G)$ contains $D(L_p^n)$; suppose $x \in D(L_p^n)$, i.e., suppose $x = D*z$ for some $z \in L_p^n$. Then $\hat{x} = \hat{D}\hat{z}$, and $\hat{G}\hat{x} = \hat{G}\hat{x} = \hat{N}\hat{D}^{-1}\hat{D}\hat{z} = \hat{N}\hat{z}$, so that $Gx \in L_p^n$, which shows that $x \in M_p(G)$. To prove the opposite containment, suppose $x \in M_p(G)$, so that $x \in L_p^n$ and $Gx \in L_p^n$. Select $\hat{P},\hat{Q}$ in $\hat{A}^{n\times n}$ such that (29) (and therefore (30)) holds. Multiplying both sides of (30) by $\hat{x}$ gives

35 $$\hat{P}\,\hat{G}\,\hat{x} + \hat{Q}\,\hat{x} = \hat{D}^{-1}\,\hat{x}$$

Let $\hat{z} = \hat{D}^{-1}\hat{x}$; then, in the time domain, (35) becomes

36 $$P*Gx + Q*x = z$$

Since x and Gx both belong to L_p^n , (36) shows that $z \in L_p^n$, so that x is of the form $D*z$ where $z \in L_p^n$. This shows that $x \in D(L_p^n)$. □

Remarks (i) Theorem (32) characterizes $M_p(G)$ for every value of $p \in [1,\infty]$.

(ii) In (34), one can use any r.c.f. $(\hat{N},\hat{D})$ of $\hat{G}(.)$.

In the case where $p = 2$, we can also characterize $M^{\perp}(G)$.

37 Theorem Let $G : L_{2e}^n$ be as in Theorem (32), and let $(\hat{N},\hat{D})$ be any r.c.f. of $\hat{G}(.)$. Then

38 $$M^{\perp}(G) = \{x \in L_2^n : \int_0^{\infty} D'(\tau)\, x(\tau+t)\, d\tau = 0 \quad \forall t \geq 0\}$$

<u>Proof</u> By Theorem (32), we have

39 $$M_2(G) = \{x \in L_2^n : x = Dz \quad \text{for some} \quad z \in L_2^n\}$$

Hence

40 $$M^{\perp}(G) = \{x \in L_2^n : \langle x, Dz\rangle = 0 \quad \text{for all} \quad z \in L_2^n\}$$
$$= \{x \in L_2^n : D^* x = 0\}$$

where $D^* : L_2^n \to L_2^n$ is the adjoint operator of D defined by

41 $$\langle x, Dz\rangle = \langle D^* x, z\rangle$$

In the present case, we have

42 $$\langle x, Dz\rangle = \int_0^{\infty} x'(t) \int_0^{t} D(t-\tau)\, z(\tau)\, d\tau\, dt$$
$$= \int_0^{\infty} z'(\tau) \int_{\tau}^{\infty} D'(t-\tau)\, x(t)\, dt\, d\tau$$

so that

43 $$(D^* x)(\tau) = \int_{\tau}^{\infty} D'(t-\tau)\, x(t)\, dt$$

Hence, by (40),

44 $$M^{\perp}(G) = \{x \in L_2^n : \int_t^{\infty} D'(\tau-t)\, x(\tau)\, d\tau = 0 \quad \forall t \geq 0\}$$
$$= \{x \in L_2^n : \int_0^{\infty} D'(\tau)\, x(\tau+t)\, d\tau = 0 \quad \forall t \geq 0\}$$

which proves (38). □

Theorems (32) and (37), though they give complete characterizations of the sets $M_p(G)$ and $M^{\perp}(G)$, are still somewhat abstract. The next result gives more concrete results, which are especially useful in Chapters 8 and 9.

45 <u>Theorem</u> Suppose $G : L_{2e}^n \to L_{2e}^n$ is as in Theorem (32), has an r.c.f. $(\hat{N},\hat{D})$, and belongs to class U . Let $s_o \in C_+^o$ be a pole of $\hat{G}(.)$. Under these conditions.

(i) if s_o is complex, let v be a vector in C^n such that $\hat{D}'(s_o)\ v = 0$. Then $x(t) = v \exp(-s_o t) + v^* \exp(-s_o^* t)$, where * denotes complex conjugate, belongs to $M^{\perp}(G)$;

(ii) if s_o is real, let v be a vector in R^n such that $\hat{D}'(s_o)\ v = 0$. Then $x(t) = v \exp(-s_o t)$ belongs to $M^{\perp}(G)$.

<u>Proof</u> First of all, if $s_o \in C_+^o$ is a pole of $\hat{G}(.)$, then by Lemma (28), $\det \hat{D}(s_o) = 0$, so that $\det \hat{D}'(s_o) = 0$ also. Hence there exists a nonzero vector v such that $\hat{D}'(s_o)\ v = 0$. Now suppose s_o is complex; then $\hat{D}^+(s_o)\ v^* = 0$. Let

46 $$x(t) = v \exp(-s_o t) + v^* \exp(-s_o^* t)$$

Clearly $x(.) \in L_2^n$, because $s_o \in C_+^o$ (i.e. Re $s_o > 0$). Also,

47 $$\int_0^\infty D'(\tau)\ x(\tau+t)\ d\tau$$

$$= \int_0^\infty D'(\tau)\ v \exp(-s_o\tau).\exp(-s_o t)\ d\tau$$

$$+ \int_0^\infty D'(\tau)\ v^* \exp(-s_o^*\tau).\exp(-s_o^* t)\ d\tau$$

$$= \hat{D}'(s_o) v \exp(-s_o t) + \hat{D}^+(s_o) v^* \exp(-s_o^* t) = 0 ,$$

$$\forall t \geq 0$$

which shows that $x \in M^{\perp}(G)$, by Theorem (37). The case where s_o is real follows similarly. □

Theorem (45) demonstrates <u>some</u> elements of the subspace $M^{\perp}(G)$, in the special case where G belongs to U and

$\hat{G}(.)$ has a pole in C_+^o. However, it does not say that <u>all</u> elements of $M^\perp(G)$ are of the form (46).

We close out this section by giving alternate characterizations of conditional gain and conditional dissipativity for a linear convolution operator.

48 <u>Lemma</u> Suppose G is as in Theorem (45). Then

49
$$\zeta_{c2}(G) = \sup_\omega \{\lambda_{max}[\hat{G}^+(j\omega)\,\hat{G}(j\omega)]\}^{1/2}$$

<u>Proof</u> The proof is entirely analogous to that of (3.1.83). In fact (3.1.84) applies, if we note that $Gx \in L_2^n$ whenever $x \in M_2(G)$. □

50 <u>Lemma</u> Suppose G is as in Theorem (45), and let $(\hat{N},\hat{D})$ be any r.c.f. of $\hat{G}(.)$. Then G is conditionally dissipative with respect to a given triplet (Q,R,S) if and only if the matrix

51
$$P(\omega) = [\hat{D}^+(j\omega)S\,\hat{N}(j\omega) + \hat{N}^+(j\omega)S'\hat{D}(j\omega)]/2 + \hat{D}^+(j\omega)R\,\hat{D}(j\omega) + \hat{N}^+(j\omega)Q\,\hat{N}(j\omega)$$

is positive semidefinite for all ω.

<u>Proof</u> By Definition (1), G is (Q,R,S) conditionally dissipative if and only if

52
$$\langle Gx,QGx\rangle + \langle x,Rx\rangle + \langle Gx,Sx\rangle \geq 0 \quad \forall x \in M_2(G)$$

By Theorem (32), $x \in M_2(G)$ if and only if $\hat{x} = \hat{D}\hat{z}$ for some $z \in L_2^n$, in which case $\hat{G}\hat{x} = \hat{N}\hat{z}$. If we abuse notation slightly by using D to denote the operator mapping z into $D*z$, and define $Nz = N*z$, then (52) is equivalent to

53
$$\langle Nz,QNz\rangle + \langle Dz,RDz\rangle + \langle Nz,SDz\rangle \geq 0\,, \quad \forall z \in L_2^n$$

i.e.,

54
$$\langle z,(N^*QN + D^*RD - D^*SN)z\rangle \geq 0 \quad \forall z \in L_2^n$$

where * denotes the adjoint operator. By Parseval's equality, (54) holds if and only if $P(\omega)$ is positive semidefinite for all ω.

NOTES AND REFERENCES

The utility of the concept of gain in studying feedback stability arises from the work of Sandberg [San. 2] and Zames [Zam. 3]. The gain calculations of Section 3.1 are taken from many sources, including the work of Sandberg [San. 3], Wu and Desoer [Wu 1] and Willems [Wil. 1]. The concept of passivity is of course well-established in circuit and system theory. The more general concept of dissipativity was introduced by Willems [Wil. 3,4,5] and exploited by Hill and Moylan [Hil. 1,2], [Moy. 2]. The concept of conditional gain is implicit in the work of Takeda and Bergen [Tak. 1], while conditional dissipativity was introduced (under a different name) by Hill and Moylan [Hil. 3], [Moy. 3]. The properties of a class U operator are identified in [Tak. 1]. The calculation of the domain and range of an unstable operator are found in [Vid. 5], while the calculation of $M^{\perp}(G)$ can be found in [Vid. 7].

CHAPTER 4: DECOMPOSITION OF LARGE-SCALE INTERCONNECTED SYSTEMS

In this chapter, we show how the study of a general large-scale interconnected system (LSIS) can be simplified by first decomposing the LSIS into its so-called strongly connected components (SCC's). By using this procedure, one never has to deal with a more complex system than the one with which he began, and in many cases, the original system is replaced by several smaller systems which <u>together</u> are simpler than the original system. The primary tool used to effect this decomposition is the theory of directed graphs, or <u>digraphs</u>, and we begin with a development of the results that we need from this theory. It should be mentioned at the outset that the following discussion is highly selective; the reader is referred to [Baa. 1] for a more complete discussion.

4.1 SOME RESULTS FROM THE THEORY OF DIRECTED GRAPHS

4.1.1 Basic Concepts

1 <u>Definition</u> A <u>directed graph</u> (or <u>digraph</u>) is a pair (V,E), where $V = \{v_1,\ldots,v_n\}$ is a finite set and E is a subset of $V \times V$. The set V is referred to as the <u>set of vertices</u>, and E is referred to as the <u>set of edges</u>. If $(v_i,v_j) \in E$, then we say that there is an edge <u>from</u> v_i <u>to</u> v_j.

Apart from the above abstract definition, we often use a pictorial representation of a digraph (V,E), whereby n points in the plane are labelled as $v_1,\ldots,v_n$, and an arc is drawn from v_i to v_j with an arrowhead directed towards v_j whenever $(v_i,v_j) \in E$.

2 <u>Example</u> Consider the digraph (V,E), where

3 $$V = \{v_1, v_2, v_3, v_4\}$$

4 $$E = \{(v_1,v_2),(v_1,v_4),(v_2,v_2),(v_2,v_3),(v_3,v_1),(v_3,v_4),(v_4,v_2)\}$$

The pictorial representation of this digraph is shown in

Figure 4.1 .

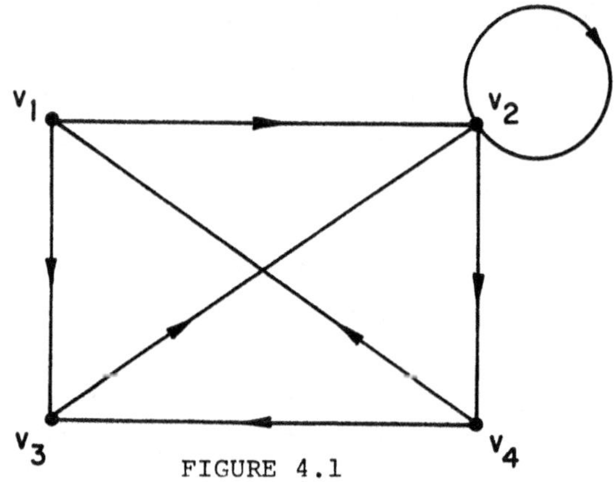

FIGURE 4.1

5 <u>Definition</u> Given a digraph (V,E), a <u>path</u> of <u>length</u> k from $v_i \in V$ to $v_j \in V$ is an ordered set $\{v_{i_0}, v_{i_1}, \ldots, v_{i_k}\}$ with $v_{i_0} \triangleq v_i$ and $v_{i_k} \triangleq v_j$, such that $(v_{i_\ell}, v_{i_{\ell+1}}) \in E$ for $\ell = 0,\ldots,k-1$.

One can state the above definition in the following equivalent manner: A path of length k from v_i to v_j is a sequence of vertices $\{v_{i_0}, v_{i_1}, \ldots, v_{i_k}\}$, with $v_{i_0} = v_i$ and $v_{i_k} = v_j$, such that there is an edge <u>from</u> v_{i_ℓ} <u>to</u> $v_{i_{\ell+1}}$ for $\ell = 0,\ldots,k-1$.

6 <u>Example</u> In the digraph of Figure 4.1, there is a path from every vertex to every other vertex.

7 <u>Definition</u> Given a digraph (V,E), we say that $v_j \in V$ <u>is reachable from</u> $v_i \in V$ (denoted by $v_i \, R \, v_j$) if there is a path from v_i to v_j. By convention, we take $v_i \, R \, v_i$ for all i, even if there is no path from v_i to itself.

Note that R defines a binary relation on V. It is obvious that R is <u>transitive</u>, i.e., $v_i \, R \, v_j$ and $v_j \, R \, v_k$ imply that $v_i \, R \, v_k$ (in other words, if there is a path from v_i to v_j and a path from v_j to v_k, then there is a path from v_i to v_k).

8 <u>Definition</u> Given a digraph (V,E), we say that a pair of vertices (v_i, v_j) is <u>strongly connected</u> if $v_i \, R \, v_j$ and

$v_j \; R \; v_i$.

If $i \neq j$, then (v_i, v_j) is strongly connected if there is a path from v_i to v_j and a path from v_j to v_i . By convention, a pair (v_i, v_i) is taken to be strongly connected even if there is no path from v_i to itself.

9 <u>Definition</u> A digraph (V,E) is <u>strongly connected</u> if every pair $(v_i, v_j) \in V \times V$ is strongly connected.

4.1.2. Testing for Strong Connectivity

We next discuss two computer algorithms for determining whether or not a given digraph is strongly connected. Note that the algorithms given here are only two of several possible ones; we select these particular algorithms because they require very little background. For a more complete discussion, the reader is referred to [Baa. 1].

The algorithms given here make use of Boolean variables, which are defined next.

10 <u>Definition</u> The <u>Boolean set</u> $(B,+,\cdot)$ is the set $B = \{0,1\}$ together with the binary operations + and $\cdot$ on B defined by

11 $$0+0 = 0, \; 0+1 = 1+0 = 1+1 = 1$$

12 $$0 \cdot 0 = 1 \cdot 0 = 0 \cdot 1 = 0, \; 1 \cdot 1 = 1$$

Those familiar with abstract algebra can verify that $(B,+,\cdot)$ is a commutative algebra. For the rest, it suffices to note that the binary operations + and $\cdot$ are associative and commutative, and that + is left - and right - distributive over $\cdot$; in other words, for all a,b,c in B, we have

13 $$a+b = b+a$$

14 $$a+(b+c) = (a+b)+c$$

15 $$a \cdot b = b \cdot a$$

16 $$a \cdot (b \cdot c) = (a \cdot b) \cdot c$$

17 $$(a+b) \cdot c = (a \cdot c)+(b \cdot c)$$

18 $$a \cdot (b+c) = (a \cdot b)+(a \cdot c)$$

Sometimes + is referred to as "or", and "$\cdot$" is referred to as "and". Also, we can speak of <u>Boolean matrices</u> over the algebra $(B,+,\cdot)$. Specifically, if $\underset{\sim}{D}$ is an array consisting of n rows and m columns such that $d_{ij} \in B$ for all i,j, we denote this by $\underset{\sim}{D} \in B^{n\times m}$. Matrix addition and multiplication are defined in the usual way; i.e., $\underset{\sim}{E} \in B^{n\times m}$, then $\underset{\sim}{F} = \underset{\sim}{D} + \underset{\sim}{E}$ also belongs to $B^{n\times m}$ and is defined by

19 $$f_{ij} = d_{ij} + e_{ij}$$

where the addition is according to (11). Similarly, if $\underset{\sim}{D} \in B^{n\times m}$ and $E \in B^{\times \ell}$, then $\underset{\sim}{F} = \underset{\sim}{D}\cdot\underset{\sim}{E}$ belongs to $B^{n\times \ell}$ and is defined by

20 $$b_{ij} = \sum_{k=1}^{m} d_{ik}\cdot e_{kj}$$

21 <u>Definition</u> Given a digraph (V,E) with $V = \{v_1,\ldots, v_n\}$, we define its <u>adjacency matrix</u> $\underset{\sim}{A} \in B^{n\times n}$ by

22 $$a_{ij} = \begin{cases} 1 & \text{if } (v_i,v_j) \in E \\ 0 & \text{if } (v_i,v_j) \notin E \end{cases}$$

23 <u>Definition</u> Given a digraph (V,E) with $V = \{v_1,\ldots, v_n\}$, we define its <u>reachability matrix</u> $\underset{\sim}{R} \in B^{n\times n}$ by

24 $$r_{ij} = \begin{cases} 1 & \text{if } v_i \; R \; v_j \\ 0 & \text{otherwise} \end{cases}$$

Note that, for any digraph, we have

25 $$r_{ii} = 1 \quad \text{for all } i$$

It is clear that a digraph is strongly connected if and only if all elements of its reachability matrix are equal to 1 . It is also clear that the adjacency matrix of a digraph can be formed by inspection. In order to develop an algorithm to test for strong-connectedness, we derive a method for computing $\underset{\sim}{R}$ given $\underset{\sim}{A}$. This is the purpose of the next several lemmas.

26 <u>Lemma</u> Consider the matrix $\underset{\sim}{A}^{\ell}$, obtained by multiplying $\underset{\sim}{A}$ by itself ℓ times (for $\ell \geq 1$) . Then

27 $$(\underset{\sim}{A}^{\ell})_{ij} = \begin{cases} 1 & \text{if there is a path of length } \ell \text{ from } v_i \text{ to } v_j \text{ .} \\ 0 & \text{otherwise} \end{cases}$$

<u>Proof</u> The proof is by induction. First, when $\ell = 1$, note that a path of length 1 from v_i to v_j is just an edge (v_i, v_j). Hence (27) is true for $\ell = 1$. Now suppose (27) is true for $\ell = k-1$. Then for $\ell = k$, we have

28 $$(\underset{\sim}{A}^{k})_{ij} = \sum_{m=1}^{n} (\underset{\sim}{A}^{k-1})_{im} \cdot a_{mj}$$

From the definitions of + and . (namely (11) and (12)), it is easy to see that $(\underset{\sim}{A}^{k})_{ij} = 1$ if and only if

29 $$\exists\ m \text{ such that } (\underset{\sim}{A}^{k-1})_{im} = 1 \text{ and } a_{mj} = 1$$

By the inductive hypothesis, (27) is true for $\ell = k-1$. Thus (29) is equivalent to

30 $\exists$ m such that there is a path of length k-1 from v_i to v_m and an edge (v_m, v_j).

Clearly (30) is equivalent to

31 There is a path of length k from v_i to v_j .

Thus $(\underset{\sim}{A}^k)_{ij} = 1$ if and only if (31) holds. This shows that (27) is true for $\ell = k$, and completes the proof by induction. □

32 **Lemma** Given a digraph, we have

33 $$\underset{\sim}{R} = \sum_{i=0}^{\infty} \underset{\sim}{A}^i$$

Proof Since $\underset{\sim}{R} = \underset{\sim}{I} + \underset{\sim}{A} + \underset{\sim}{A}^2 + \ldots$ (where $\underset{\sim}{I}$ denotes the identity matrix), we see that $r_{ii} = 1$ for all i. For $i \neq j$, we see that $r_{ij} = 1$ if and only if $(\underset{\sim}{A}^{\ell})_{ij} = 1$ for some $\ell \geq 1$, i.e., if and only if there is a path (of some length ℓ) from v_i to v_j. □

34 **Lemma** Given a digraph with n vertices, we have

35 $$\underset{\sim}{R} = \sum_{i=0}^{n-1} \underset{\sim}{A}^i$$

Proof The lemma is established if we can show that, whenever there is a path from v_i to $v_j \neq v_i$, there is a path of length at most $n-1$ from v_i to v_j. Accordingly, suppose there is a path of length k from v_i to v_j, and let $\{v_{i_0} \triangleq v_i, \ldots, v_{i_{k-1}}, v_{i_k} \triangleq v_j\}$ be the sequence of vertices defining this path. If $k \geq n$, then the list of $k+1$ vertices $(v_{i_0}, \ldots, v_{i_k})$ contains at least two elements that are the same. Suppose $v_{i_\ell} = v_{i_m}$ where $0 \leq \ell < m \leq k$. Then clearly the sequence of vertices $\{v_{i_0}, \ldots, v_{i_\ell}, v_{i_{m+1}}, \ldots, v_{i_k}\}$ also defines a path from v_i to v_j, and it has length $k-(m-\ell)$. If this latter quantity is again larger than $n-1$, we can repeat the argument until we find a path of length at most $n-1$ from v_i to v_j. □

36 **Lemma** Given a digraph with n vertices, we have

37 $$\underset{\sim}{R} = (\underset{\sim}{I} + \underset{\sim}{A})^s \quad \text{for every } s \geq n-1$$

Proof First, from the proof of Lemma (35), it is clear that in fact

38 $$R = \sum_{i=0}^{s} A^i \quad \text{for every} \quad s \geq n-1$$

We now claim that

39 $$\sum_{i=0}^{s} A^i = (I + A)^s \quad \text{for every} \quad s \geq 0$$

Clearly, if (39) can be established, then the lemma is proved. We prove (39) by induction. First, (39) is true for $s = 0$, since both sides equal I in this case. Next, suppose (39) is true for $s = \ell$, and observe that $M + M = M$ for all matrices M. Thus

40 $$(I + A)^{\ell+1} = (I + A)^{\ell} (I + A) = \left(\sum_{i=0}^{\ell} A^i\right) (I + A)$$
$$= \sum_{i=0}^{\ell} A^i + \sum_{i=0}^{\ell} A^{i+1}$$
$$= \sum_{i=0}^{\ell+1} A^i$$

This proves the lemma. □

The discussion up to now suggests the following algorithm for computing R given A.

41 <u>Algorithm for computing The Reachability Matrix R</u>

<u>Step 1</u> Given A, let $M = I + A$.

<u>Step 2</u> Select an integer t such that $2^t \geq n-1$.

<u>Step 3</u> Calculate R as M^{2^t}.

Note that step 3 can be accomplished by t matrix multiplications ($M.M = M^2$, $M^2.M^2 = M^4$, etc.). Sometimes fewer than t matrix multiplications are required. For instance, if $M^{2^s} . M^{2^s} = M^{2^s}$ for some $s < t$, then clearly $R = M$. Thus if the squaring process $(\underset{\sim}{M}^{2^s} \cdot \underset{\sim}{M}^{2^s} = \underset{\sim}{M}^{2^{s+1}})$ does not result in a new matrix at some stage, then that is the matrix $\underset{\sim}{R}$.

It is obvious that the above algorithm requires $\log_2 n$ matrix multiplications. Now, it can be shown that two Boolean matrices of order $n \times n$ can be multiplied using $(n^2/\log_2 n)$ row additions (i.e., "or" operations) [Baa. 1, pp. 226-231]. Thus, computation of $\underset{\sim}{R}$ requires n^2 row additions.

We now present another algorithm, known as Warshall's algorithm, for computing $\underset{\sim}{R}$. This algorithm also requires n^2 row additions, and is given without proof. Note that we use the standard notation "$\underset{\sim}{A}_1 \leftarrow \underset{\sim}{A}_2$" to mean "$\underset{\sim}{A}_2$ replaces $\underset{\sim}{A}_1$." The details of this algorithm can be found in [Baa. 1, pp. 222-223].

42 <u>Warshall's Algorithm for computing $\underset{\sim}{R}$</u>

<u>Step 1</u> $\underset{\sim}{R} \leftarrow \underset{\sim}{A}$

<u>Step 2</u> For $k \leftarrow 1$ to n, do

For $i \leftarrow 1$ to n, do

If $r_{ik} = 1$ then $\underset{\sim}{R}^i = \underset{\sim}{R}^i + \underset{\sim}{R}^k$

($\underset{\sim}{R}^i$ = i-th row of $\underset{\sim}{R}$)

<u>Step 3</u> $\underset{\sim}{R} \leftarrow \underset{\sim}{R}+\underset{\sim}{I}$

43 <u>Example</u> Consider the digraph of Figure 4.2 . Its adjacency matrix is

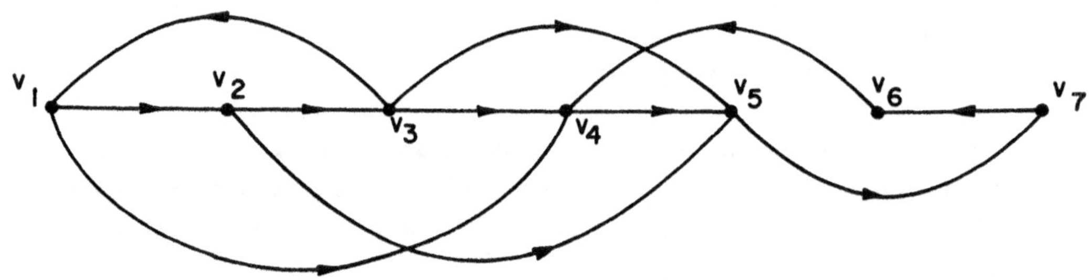

FIGURE 4.2

44 $$\underset{\sim}{A} = \begin{bmatrix} 0 & 1 & 0 & 1 & 0 & 0 & 0 \\ 0 & 0 & 1 & 0 & 1 & 0 & 0 \\ 1 & 0 & 0 & 1 & 1 & 0 & 0 \\ 0 & 0 & 0 & 0 & 1 & 0 & 0 \\ 0 & 0 & 0 & 0 & 0 & 0 & 1 \\ 0 & 0 & 0 & 1 & 0 & 0 & 0 \\ 0 & 0 & 0 & 0 & 0 & 1 & 0 \end{bmatrix}$$

Using either Algorithm (41) or Algorithm (42), we can compute $\underset{\sim}{R}$ as

45 $$\underset{\sim}{R} = \begin{bmatrix} 1 & 1 & 1 & 1 & 1 & 1 & 1 \\ 1 & 1 & 1 & 1 & 1 & 1 & 1 \\ 1 & 1 & 1 & 1 & 1 & 1 & 1 \\ 0 & 0 & 0 & 1 & 1 & 1 & 1 \\ 0 & 0 & 0 & 1 & 1 & 1 & 1 \\ 0 & 0 & 0 & 1 & 1 & 1 & 1 \\ 0 & 0 & 0 & 1 & 1 & 1 & 1 \end{bmatrix}$$

Since $\underset{\sim}{R}$ does not contain all 1's, the digraph is not strongly connected.

4.1.3. Decomposition into Strongly Connected Components

Suppose that a given digraph is not strongly connected. Can we deduce some further information about its structure? The answer is provided by identifying the so-called strongly connected components (SCC's) of the digraph.

46 Definition Given a digraph (V,E), the binary relation S on V is defined by $v_i \; S \; v_j$ if and only if the pair (v_i,v_j) is strongly connected (i.e., $v_i \; R \; v_j$ and $v_j \; R \; v_i$).

It is easy to show that S is in fact an equivalence relation on V. Since S is an equivalence relation we can say (without ambiguity) that "a set of vertices is strongly connected," when what we mean is that the vertices are pairwise strongly connected.

Since S is an equivalence relation on V, we can partition V into its equivalence classes under S. Clearly a digraph (V,E) is strongly connected if and only if V itself is an equivalence class under S. Otherwise, the equivalence classes of V under S can be easily identified using the reachability matrix $\underset{\sim}{R}$.

48 <u>Lemma</u> Let (V,E) be a given digraph and let $\underset{\sim}{R}$ denote its reachability matrix. Then $v_i \; S \; v_j$ if and only if $\underset{\sim}{R}^i = \underset{\sim}{R}^j$, where $\underset{\sim}{R}^i$ denotes the j-th row of $\underset{\sim}{R}$.

<u>Proof</u> "if" Suppose $\underset{\sim}{R}^i = \underset{\sim}{R}^j$. By definition, $r_{ii} = 1$, so the hypothesis implies that $r_{ji} = 1$, i.e., $v_j \; R \; v_i$. Similarly, since $r_{ij} = r_{jj} = 1$, we have $v_i \; R \; v_j$. Hence $v_i \; S \; v_j$.

"only if" Suppose $v_i \; S \; v_j$, i.e., $v_i \; R \; v_j$ and $v_j \; R \; v_i$. Now, whenever $r_{jk} = 1$ for some k, we also have $r_{ik} = 1$; this is because $r_{jk} = 1$ implies $v_j \; R \; v_k$, which together with $v_i \; R \; v_j$ implies that $v_i \; R \; v_k$, i.e., $r_{ik} = 1$. By a symmetrical argument, $r_{ik} = 1$ for some k implies $r_{jk} = 1$. In other words, $\underset{\sim}{R}^i = \underset{\sim}{R}^j$. □

49 <u>Example</u> Consider again the digraph of Figure 4.2. From (45), we immediately see that the equivalence classes under S are $\{v_1,v_2,v_3\}$ and $\{v_4,v_5,v_6,v_7\}$.

In general, once R is computed using either of the Algorithms (41) or (42), it is easy to determine the equivalence classes under S by comparing each row of $\underset{\sim}{R}$ with its succeeding rows. This comparison is called "string matching," and can be efficiently carried out by computer; see [Baa. 1, Ch. 4].

Suppose we partition V into its equivalence classes under S, and label them as $V_1,\ldots,V_k$. We now show that we can order these equivalence classes in such a way that if $v_a \in V_i$, $v_b \in V_j$ and $i < j$, then $(v_b,v_a) \notin E$. In order to do this, we need some further concepts.

50 <u>Definition</u> Given a digraph (V,E), a <u>cycle</u> is a path of length greater than one from a vertex to itself. An edge of the

form (v_i, v_i) is called a <u>self-loop</u> and is not considered to be a cycle. (Note that (v_i, v_i) is a path of length one from v_i to itself). The digraph (V,E) is <u>acyclic</u> if it is does not contain any cycles (it may however contain self-loops).

51 <u>Definition</u> Given a digraph (V,E), we say that $v_i \in V$ is a <u>predecessor</u> of $v_j \in V$, and v_j is a successor of v_i, if $v_i \neq v_j$, and $(v_i, v_j) \in E$.

52 <u>Lemma</u> Suppose the digraph (V,E) is acyclic. Then there exists a vertex in V that does not have a predecessor.

<u>Proof</u> Assume the contrary, i.e., suppose every vertex has a predecessor, and construct a sequence in V as follows: Select $v_{i_0} \in V$ arbitrarily, and select $v_{i_{k+1}}$ to be a predecessor of v_{i_k} for $k \geq 0$. By assumption, this sequence can be constructed indefinitely, and since V contains only a finite number of elements, an element of V must occur twice in the sequence. In other words we can construct a sequence $\{v_{i_\ell}, v_{i_{\ell+1}}, \ldots, v_{i_{\ell+m}} = v_{i_\ell}\}$ such that $v_{i_{k+1}}$ is a predecessor of v_{i_k} $\forall k$. Clearly $\{v_{i_{\ell+m}}, v_{i_{\ell+m-1}}, \ldots, v_{i_{\ell+1}}, v_{i_\ell} = v_{i_{\ell+m}}\}$ is a cycle in (V,E), which contradicts the hypothesis that (V,E) is acyclic. Hence the original assumption is false. □

53 <u>Proposition</u> Let $\{v_{i_1}, \ldots, v_{i_\ell} = v_{i_1}\}$ be a cycle in a digraph (V,E). Then the set of vertices $\{v_{i_1}, \ldots, v_{i_{\ell-1}}\}$ is strongly connected.

The proof is obvious.

54 <u>Definition</u> Given a digraph (V,E), let $V_1, \ldots, V_k$ denote the equivalence classes of V under S (strong connectivity). Then the reduced digraph $(\bar{V}, \bar{E})$ is defined as follows: The vertex set $\bar{V} = \{V_1, \ldots, V_k\}$, and the edge set $\bar{E}$ contains an edge (V_i, V_j) if and only if there exist $v_a \in V_i$, $v_b \in V_j$ such that $(v_a, v_b) \in E$.

The reduced digraph has a very simple interpretation.

Suppose we modify the original digraph (V,E) by collapsing all vertices in V_i into a single vertex, for $i = 1,\ldots,k$. The resulting digraph is the reduced digraph. Note that if (V,E) is strongly connected, then its reduced digraph consists of a single vertex and a self-loop.

55 <u>Lemma</u> For any digraph (V,E), its reduced digraph is acyclic.

<u>Proof</u> Assume the contrary, namely that $(V_{i_1}, V_{i_2}, \ldots, V_{i_m}, V_{i_{m+1}} = V_{i_1}\}$ is a cycle in $(\bar{V},\bar{E})$. This means that $\bar{E}$ contains the edges $(V_{i_2},V_{i_1}),\ldots,(V_{i_m},V_{i_1})$. By the definition of $\bar{E}$, this implies that there exist vertices $v_{i_j}^{(1)}$, $v_{i_j}^{(2)}$ in V_{i_j}, $j = 1,\ldots,m$, having the following property. The original edge set E contains the edges $(v_{i_1}^{(2)}, v_{i_2}^{(1)}), (v_{i_2}^{(2)}, v_{i_3}^{(1)}),\ldots, (v_{i_{m-1}}^{(2)}, v_{i_m}^{(1)}), (v_{i_m}^{(2)}, v_{i_1}^{(1)})$. Now, since V_{i_j} is an equivalence class under S, there is a path from $v_{i_j}^{(1)}$ to $v_{i_j}^{(2)}$, for $j = 1,\ldots,k$. So what we have shown is that there is a cycle in the original digraph (V,E) containing the vertices $\{v_{i_1}^{(1)}, v_{i_1}^{(2)}, \ldots, v_{i_m}^{(1)}, v_{i_m}^{(2)}\}$. By Proposition (53), this implies that all these vertices are strongly connected. However, this is a contradiction, because $\{v_{i_1}^{(1)}, v_{i_1}^{(2)}\}$ belongs to a distinct equivalence class from $\{v_{i_2}^{(1)}, v_{i_2}^{(2)}\}$ (for example). This contradiction shows that our original assumption is false, and that the reduced digraph is acyclic. □

We can now state a procedure for renumbering the equivalence classes $V_1,\ldots,V_k$ in the manner described before Definition (50). Given a digraph (V,E), first construct its reduced digraph $(\bar{V},\bar{E})$. Now, identify all vertices in $\bar{V}$ that do not have predecessors, and label this set as $\bar{V}_1$. Renumber the vertices in $\bar{V}_1$ as $W_1,W_2,\ldots,W_{n_1}$. Next, remove from $(\bar{V},\bar{E})$ all vertices in $\bar{V}_1$ and all edges leaving vertices in $\bar{V}_1$ (i.e., all edges of the form $(V_i,.)$ with $V_i \in \bar{V}_1$). The resulting digraph is again acyclic. Identify its collection of vertices that do not have predecessors, and label it as $\bar{V}_2$. It is easy to see that if $V_i \in \bar{V}_2$, then its predecessor must belong to $\bar{V}_1$. Renumber the vertices in $\bar{V}_2$ as $W_{n_{1+1}},\ldots,W_{n_2}$. Repeat this procedure until all vertices in $\bar{V}$ are exhausted. At this stage, we have renumbered the vertices (i.e., equivalence classes) $V_1,\ldots,V_k$ as $W_1,\ldots W_k$ in such a way that all predecessors of W_i lie among $\{W_1,\ldots,W_{i-1}\}$, for $i = 2,\ldots,k$. To avoid a proliferation of symbols, let $V_1,\ldots,V_k$ now denote the vertices of $\bar{V}$ numbered in the proper order, i.e., such that all predecessors of V_i lie among $\{V_1,\ldots,V_{i-1}\}$, for $i = 2,..,k$. This means that if $i < j$, then $(V_j,V_i) \notin \bar{E}$. With reference to the <u>original</u> digraph (V,E), this means that if $v_a \in V_j$ and $i < j$, then $(v_b,v_a) \notin E$ (this follows form the definition of $\bar{E}$). Thus we have proved the following

56 <u>Theorem</u> Given a digraph (V,E), one can partition the vertex set V as $V_1,\ldots,V_k$ in such a way that

(i) $V_1,\ldots,V_k$ are the equivalence classes of V under strong connectedness;

(ii) if $v_a \in V_i$, $v_b \in V_j$, and $(v_b,v_a) \in E$, then $i \geq j$.

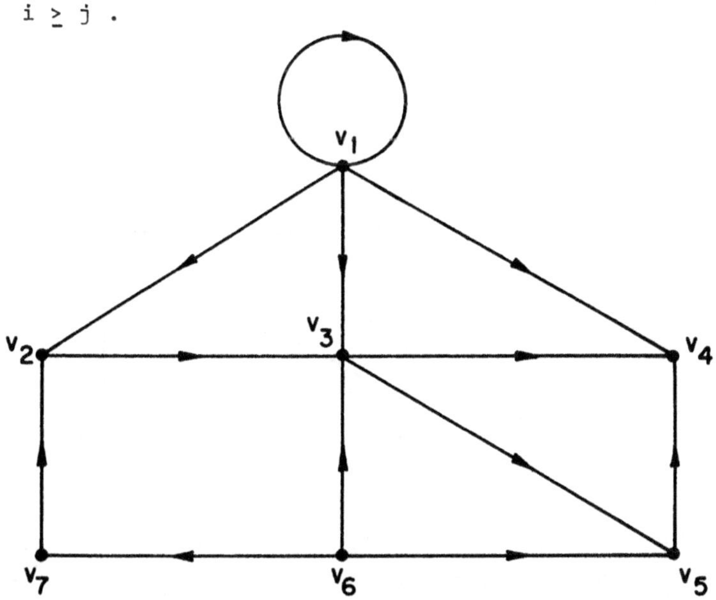

FIGURE 4.3

57 <u>Example</u> Consider the digraph of Figure 4.3, whose adjacency matrix is

58a
$$\underset{\sim}{A} = \begin{bmatrix} 1 & 1 & 1 & 1 & 0 & 0 & 0 \\ 0 & 0 & 1 & 0 & 0 & 0 & 0 \\ 0 & 0 & 0 & 1 & 1 & 0 & 0 \\ 0 & 0 & 0 & 0 & 0 & 0 & 0 \\ 0 & 0 & 0 & 1 & 0 & 0 & 1 \\ 0 & 0 & 1 & 0 & 1 & 0 & 1 \\ 0 & 1 & 0 & 0 & 0 & 0 & 0 \end{bmatrix}$$

Using either Algorithm (41) or (42), we can compute that

58b
$$\underset{\sim}{R} = \begin{bmatrix} 1 & 1 & 1 & 1 & 1 & 0 & 1 \\ 0 & 1 & 1 & 1 & 1 & 0 & 1 \\ 0 & 1 & 1 & 1 & 1 & 0 & 1 \\ 0 & 0 & 0 & 1 & 0 & 0 & 0 \\ 0 & 1 & 1 & 1 & 1 & 0 & 1 \\ 0 & 1 & 1 & 1 & 1 & 1 & 1 \\ 0 & 1 & 1 & 1 & 1 & 0 & 1 \end{bmatrix}$$

Using Lemma (48), we can determine that the equivalence classes of vertices are

59
$$V_1 = \{v_1\},\ V_2 = \{v_2, v_3, v_5, v_7\},\ V_3 = \{v_4\},\ V_4 = \{v_6\}$$

Hence the reduced digraph is as shown in Figure 4.4 .

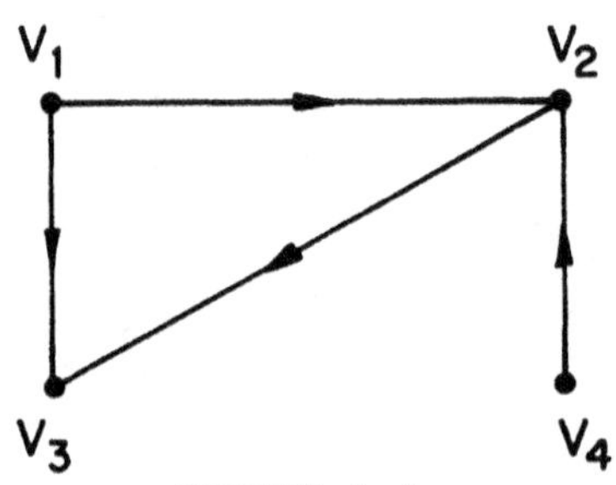

FIGURE 4.4

This digraph is acyclic, as expected. Further, the vertices V_1 and V_4 do not have predecessors. So we let $W_1 = V_1$,

$W_2 = V_4$. If we remove these vertices, as well as the edges leaving these vertices, we get the digraph shown in Figure 4.5 . Clearly V_2 does not have a predecessor, so we let $W_3 = V_2$.

FIGURE 4.5

Finally, $W_4 = V_3$. Hence the equivalence classes, numbered in the appropriate order, are

60a $$V_1 = \{v_1\},\ V_2 = \{v_6\},\ V_3 = \{v_2,v_3,v_5,v_7\},\ V_4 = \{v_4\}$$

Note that this ordering is not unique; for instance, we can also take

60b $$V_1 = \{v_6\},\ V_2 = \{v_1\},\ V_3 = \{v_2,v_3,v_5,v_7\},\ V_4 = \{v_4\}$$

We close out this subsection by formally introducing concepts of a strongly connected component and an interconnecting subgraph.

61 Definition Given a digraph (V,E), let $V_1,\ldots,V_k$ denote the equivalence classes of V under S, ordered in such a way that $(v_b,v_a) \notin E$ whenever $v_a \in V_i, V_b \in V_j$, and $i < j$. Then the digraph $(v_i, (V_i \times V_i) \cap E)$ is called the i-th strongly connected component (SCC), for $i = 1,\ldots,k$. The digraph $(V_i \cup V_j\ ,\ (V_i \times V_j) \cap E)$ is called the ij-th inter-connecting subgraph (IS), for $1 \le i \le j \le k$.

63 Example Consider once again the digraph of Figure 4.3 . Its strongly connected components are

63a $$SCC1 = (\{v_1\},\ (v_1,v_1))$$

63b $$SCC2 = (\{v_6\},\ \emptyset)$$

63c $$SCC3 = (\{v_2,v_3,v_5,v_7\},\ (v_2,v_3),\ (v_3,v_5),\ (v_5,v_7),\ (v_7,v_2))$$

63d SCC4 = $(\{v_4\}, \emptyset)$

where $\emptyset$ denotes the empty set. Its interconnecting subgraphs are

64a IS12 = $(\{v_1,v_6\}, \emptyset)$

64b IS13 = $(\{v_1,v_2,v_3,v_5,v_7\}, (v_1,v_2), (v_1,v_3))$

64c IS14 = $(\{v_1,v_4\}, (v_1,v_4))$

64d IS23 = $(\{v_6,v_2,v_3,v_5,v_7\}, (v_6,v_3), (v_6,v_5), (v_6,v_7))$

64e IS24 = $(\{v_6,v_4\}, \emptyset)$

64f IS34 = $(\{v_2,v_3,v_5,v_7,v_4\}, (v_3,v_4), (v_5,v_4))$

The important point to note is that the union of all the SCC's is a subgraph of (V,E); that is, the vertex set of the union is V, and its edge set is a subset of E . Unless (V,E) is strongly connected, the union of all the SCC's of (V,E) is a proper subgraph of (V,E), in that its edge set is a proper subset of E . For instance, for the digraph of Figure 4.3, the union of the four SCC's has only five edges. Thus, in general, the problem of analyzing all of the SCC's is still simpler than analyzing the original digraph.

4.1.4. Directed Trees

65 Definition Given a strongly connected digraph (V,E), we define a directed tree of (V,E) to be a maximal subgraph of (V,E) that contains neither cycles nor self-loops.

In other words, a directed tree of (V,E) is a subgraph (V,E_t) that contains neither cycles nor self-loops, with the additional property that the inclusion of any edge from $E-E_t$ creates either a cycle or a self-loop.

66 Example If we remove the edges (v_3,v_1), (v_4,v_1) and (v_5,v_7) from the digraph of Figure 4.2 , we get a directed tree. Alternatively, we can remove (v_1,v_2) and (v_7,v_6) and get

another directed tree.

67 <u>Lemma</u> Given a strongly connected digraph (V,E), let (V,E_s) be a subgraph of (V,E) that contains neither cycles nor self-loops. Then there is a superset E_t of E_s such that (V,E_t) is a directed tree of (V,E).

This lemma states that any subgraph that contains neither cycles nor self-loops can be imbedded in a directed tree. The proof is obvious and is omitted. This lemma is needed to prove the well-posedness results of Chapter 5 .

4.2 DECOMPOSITION INTO STRONGLY CONNECTED SUBSYSTEMS

In this section, we show how a given large-scale interconnected system (LSIS) can be decomposed into a hierarchical arrangement of subsystems, whereby each subsystem interacts only with those "below" it. The advantages of carrying out such a decomposition are discussed in Section 4.3. In what follows, we discuss in turn each of alternative system descriptions (2.2.1) and (2.2.18), beginning with (2.2.1).

Consider a large-scale interconnected system described by

1a $$e_i = u_i = \sum_{j=1}^{m} H_{ij} y_j$$

1b $$y_i = G_i e_i$$

$$\left.\right\} \quad i = 1,\ldots,m$$

where u_i, e_i, y_i belong to $L_{pe}^{n_i}$ for some fixed $p \in [1,\infty]$ and some positive integer n_i , $G_i : L_{pe}^{n_i} \to L_{pe}^{n_i}$, and $H_{ij} : L_{pe}^{n_j} \to L_{pe}^{n_i}$. Given the system (1), we associate with it a digraph constructed as follows: Label m vertices as $v_1,\ldots,v_m$, and draw an edge (v_j,v_i) if $H_{ij} \neq 0$. If the resulting digraph is not connected, then it means that (1) actually represents a collection of disjoint subsystems that do not interact with each other. In this case each connected component can be analyzed separately. Hence we can safely assume that the digraph associated with (1) (referred to hereafter as the <u>system digraph</u>) is connected.

If the system digraph is also strongly connected, then there is nothing further to be done. If the system digraph is not strongly connected, then identify the SCC's and number the vertices as indicated in Section 4.1. Once this is done, we have a renumbering and partitioning of the vertex set $V = \{v_1,\dots,v_m\}$ into $V_i = \{v_{n_{i-1}+1},\dots,v_{n_i}\}$, $i = 1,\dots,k$, in such a way that no edge (v_a,v_b) exists whenever $v_a \in V_i$, $v_b \in V_j$, and $i > j$. Now, we partition the system equations (1) in a corresponding manner. Define

$$\left.\begin{aligned} &2 \quad x_i = (u_j,\ v_j \in V_i) = (u_{n_{i-1}+1},\dots,u_{n_i}) \\ &3 \quad z_i = (y_j,\ v_j \in V_i) \\ &4 \quad d_i = (e_j,\ v_j \in V_i) \\ &5 \quad F_i = \text{Diag}\,(G_j,\ j \in V_i) \\ &6 \quad R_{ij} = (H_{rs},\ v_r \in V_i,\ v_s \in V_j) \end{aligned}\right\} \quad i = 1,\dots,k$$

Because of the renumbering, we have that $R_{ij} = 0$ whenever $i < j$. This is the objective of the decomposition into SCC's.

With the above definitions, the system equations (1) can be equivalently expressed as

7a $$d_i = x_i - R_{ii}\, z_i - \sum_{j=1}^{i-1} R_{ij}\, z_j$$

7b $$z_i = F_i\, d_i$$

where $x_i, d_i, z_i \in L_{pe}^{\nu_i}$ for some positive integer ν_i, $F_i : L_{pe}^{\nu_i} \to L_{pe}^{\nu_i}$, and $R_{ij} : L_{pe}^{\nu_j} \to L_{pe}^{\nu_i}$. Let us define

8a $$d_i = x_i - R_{ii}\, z_i$$

8b $$z_i = F_i\, d_i$$

as (S_i), or the i-th isolated subsystem. Then the LSIS (1) (or equivalently (7)) can be viewed as an interconnection of the k isolated subsystems (S_1) thru (S_k) but with the additional property that (S_i) does not interact with (S_j) if $i > j$. Thus the isolated subsystems are arranged in a hierarchy, whereby each system interacts only with those having a higher subscript. The advantage of such an arrangement is that the well-posedness and stability of the overall system can be deduced by studying the isolated subsystems alone. In this connection, it is important to note that the union of the isolated subsystems is in general simpler than the original system, because all the interconnection operators R_{ij} have been omitted. Therefore, if one tries this approach on a given system, then the worst that can happen is that the system digraph is strongly connected, in which case one has not lost anything. On the other hand, if the system digraph is not strongly connected, then considerable savings in complexity can result.

9 Example Consider a system of the form (1), where $m = 5$ (i.e., 5 subsystems), and the following interconnection operators are nonzero: $H_{21}, H_{25}, H_{32}, H_{42}, H_{43}, H_{51}, H_{53}$. The remaining H_{ij}, $i \neq j$ are assumed to be zero. (Note that we need not bother about operators of the form H_{ii}, because they represent self-loops in the system digraph, and therefore do not enter into the determination of the SCC's). The system digraph is shown in Figure 4.6. One can easily verify that there are three SCC's,

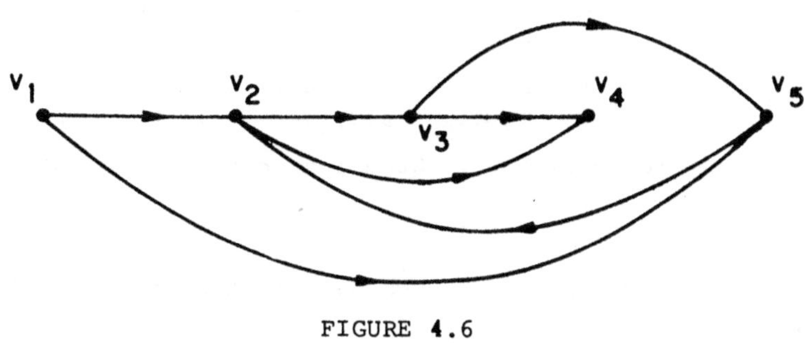

FIGURE 4.6

and that their vertex sets (arranged in the proper order) are $V_1 = \{v_i\}$, $V_2 = \{v_2, v_3, v_5\}$, $V_3 = \{v_4\}$. Accordingly, we define

10 $$x_1 = u_1, \; x_2 = [u_2, u_3, u_5]' \; , \; x_3 = u_4$$

11 $$d_1 = e_1, \; d_2 = [e_2, e_3, e_5]' \; , \; d_3 = e_4$$

12 $$z_1 = y_1, \; z_2 = [y_2, y_3, y_5]' \; , \; z_3 = y_4$$

13 $$F_1 = G_1 \; , \; F_2 = \begin{bmatrix} G_2 & 0 & 0 \\ 0 & G_3 & 0 \\ 0 & 0 & G_5 \end{bmatrix} \; , \; F_3 = G_4$$

14a $$R_{11} = H_{11} \; , \; R_{12} = H_{12} \; , \; R_{13} = 0$$

14b $$R_{22} = \begin{bmatrix} H_{22} & H_{23} & 0 \\ 0 & H_{33} & H_{35} \\ H_{52} & 0 & H_{55} \end{bmatrix} , \; R_{23} = [H_{24} \; H_{34} \; 0], \; R_{33} = H_{33}$$

The new system description is now given by (7).

It is clear that the proposed decomposition is a "structural" decomposition, in the sense that the only information that is used is whether H_{ij} is zero or nonzero. Sometimes, this can cause unnecessarily conservative results. For instance, in Example (9) above, if H_{41} were to be nonzero, then one can easily check that the system digraph would be strongly connected, and no simplification would result by this procedure. On the other hand, H_{41} might be very "small" and it may therefore be possible to safely ignore it. Such issues are discussed in Chapter 6.

Now we consider a system described by the set of equations

15 $$e_i = u_i - \sum_{j=1}^{m} S_{ij} \, e_j \; , \quad i = 1, \ldots, m$$

where $e_i, u_i \in L_{pe}^{n_i}$ for some fixed $p \in [1,\infty]$ and some positive integer n_i , and $S_{ij} : L_{pe}^{n_j} \to L_{pe}^{n_i}$. The system digraph associated with the set of equations (15) is constructed in a manner entirely analogous to before. We label m vertices as $v_1,\ldots,v_m$ and draw an edge (v_j,v_i) if $S_{ij} \neq 0$. Once this is done, we find the strongly connected components of the system digraph, and renumber the vertices as $V_i = \{v_{n_{i-1}+1},\ldots,v_{n_i}\}$ in such a way that no edge (v_a,v_b) exists whenever $v_a \in V_i$, $v_b \in V_j$, and $i > j$. We then define

16 $$x_i = (u_j \ , \ v_j \in V_i)$$

17 $$d_i = (e_j \ , \ v_j \in V_i)$$

18 $$z_{ij} = (S_{k\ell}\, e_\ell \ , \ v_k \in V_i \ , \ v_\ell \in V_j)$$

19 $$R_{ij} = (S_{k\ell} \ , \ v_k \in V_i \ , \ v_\ell \in V_j)$$

where it is important to note that the $m_i m_j$ components of z_{ij} are arranged in an $m_i m_j \times 1$ vector, whereas the $m_i m_j$ components of R_{ij} are arranged in an $m_i \times m_j$ matrix. With these definitions, the system equations can be rewritten as

20 $$d_i = x_i - \sum_{j=1}^{i} R_{ij}\, d_j$$

because, by construction, $R_{ij} = 0$ whenever $i < j$. We refer to the system

21 $$d_i = x_i - R_{ii}\, d_i$$

as the i-th isolated subsystem.

We can further modify the system description (21). From (18) we see that

22 $$z_{ij} = \tilde{R}_{ij}\, d_j$$

where

23 $$\tilde{R}_{ij} = \text{Diag [Columns of } R_{ij}]$$

Moreover, we have

24 $$R_{ij} = K_{ij}\,\tilde{R}_{ij}$$

where the operator K_{ij} is represented by the $m_i \times m_i m_j$ matrix

25 $$K_{ij} = [I_{m_i} | \dots | I_{m_i}]$$

and I_{m_i} denotes the identity matrix of dimension $m_i \times m_i$. If $(I+R_{ii})^{-1}$ exists for all i, we can express the system equations (20) solely in terms of the z_{ij}'s, as follows:

26 $$\begin{aligned} z_{ij} &= \tilde{R}_{ij}\,d_j = \tilde{R}_{ij}\,(I+R_{jj})^{-1}\,(x_j - \sum_{k=1}^{j-1} R_{jk}\,d_k) \\ &= \tilde{R}_{ij}(I+R_{jj})^{-1}\,(x_j - \sum_{k=1}^{j-1} K_{jk}\,\tilde{R}_{jk}\,d_k) \\ &= \tilde{R}_{ij}(I+R_{jj})^{-1}\,(x_j - \sum_{k=1}^{j-1} K_{jk}\,z_{jk}) \end{aligned}$$

Hence the final form of the system equations is

27 $$z_{ij} = \tilde{R}_{ij}(I+R_{jj})^{-1}\,(x_j - \sum_{k=1}^{j-1} K_{jk}\,z_{jk})$$

where it is important to observe that the operator K_{jk} is represented by a constant matrix.

28 Example Consider a system described by (15), where the following operators are nonzero: S_{11}, S_{13}, S_{21}, S_{22}, S_{32}, S_{41}, S_{43}, S_{46}, S_{52}, S_{53}, S_{54}, S_{67}, S_{75}, S_{77} . The associated system digraph is shown in Figure 4.7. (Note that the self-loops corresponding to the operators S_{11}, S_{22}, and S_{77} are not shown in Figure 4.7 because self-loops do not figure in the determination of the strongly connected subsystems).

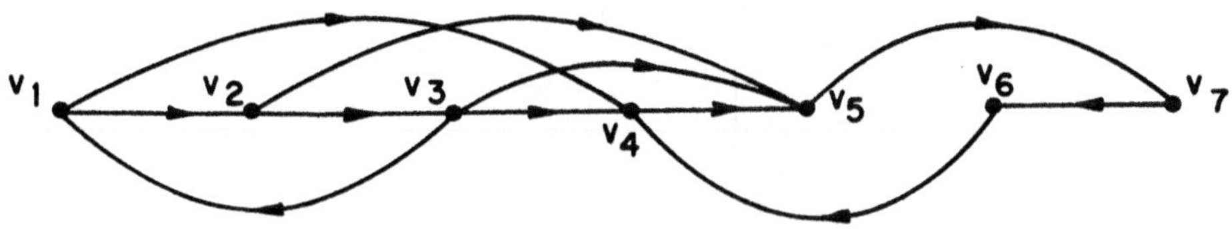

FIGURE 4.7

From Figure 4.7, we see that the equivalence classes of vertices, in the appropriate order, are $V_1 = \{v_1, v_2, v_3\}$, $V_2 = \{v_4, v_5, v_6, v_7\}$. Accordingly, we define

29 $$d_1 = \begin{bmatrix} e_1 \\ e_2 \\ e_3 \end{bmatrix}, \quad d_2 = \begin{bmatrix} e_4 \\ e_5 \\ e_6 \\ e_7 \end{bmatrix}$$

30 $$R_{11} = \begin{bmatrix} s_{11} & 0 & s_{13} \\ s_{21} & s_{22} & 0 \\ 0 & s_{32} & 0 \end{bmatrix}, \quad R_{21} = \begin{bmatrix} s_{41} & 0 & s_{43} \\ 0 & s_{52} & s_{53} \\ 0 & 0 & 0 \\ 0 & 0 & 0 \end{bmatrix}$$

$$R_{22} = \begin{bmatrix} 0 & 0 & s_{46} & 0 \\ s_{54} & 0 & 0 & 0 \\ 0 & 0 & 0 & s_{67} \\ 0 & s_{75} & 0 & s_{77} \end{bmatrix}$$

and of course $R_{12} = 0$. Moreover, we have

31 $$z_{11} = \begin{bmatrix} S_{11}\, e_1 \\ S_{21}\, e_1 \\ S_{31}\, e_1 \\ S_{12}\, e_2 \\ S_{22}\, e_2 \\ S_{32}\, e_2 \\ S_{13}\, e_3 \\ S_{23}\, e_3 \\ S_{33}\, e_3 \end{bmatrix}$$

and z_{12}, z_{21}, z_{22} are similarly defined. In (31), we display e.g. $S_{31}\, e_1$ instead of 0 (note that $\tilde{S}_{31} = 0$) to make the pattern clearer. Next, the operator $\tilde{R}_{11}$ is defined by

32 $$\tilde{R}_{11} = \begin{bmatrix} S_{11} & 0 & 0 \\ S_{21} & 0 & 0 \\ S_{31} & 0 & 0 \\ 0 & S_{12} & 0 \\ 0 & S_{22} & 0 \\ 0 & S_{32} & 0 \\ 0 & 0 & S_{13} \\ 0 & 0 & S_{23} \\ 0 & 0 & S_{33} \end{bmatrix}$$

while $\tilde{R}_{21}, \tilde{R}_{22}$ are similarly defined. Note that $\tilde{R}_{11} = K_{11}\, R_{11}$, where

33 $$\underset{\sim}{K}_{11} = \begin{bmatrix} 1 & 0 & 0 & 1 & 0 & 0 & 1 & 0 & 0 \\ 0 & 1 & 0 & 0 & 1 & 0 & 0 & 1 & 0 \\ 0 & 0 & 1 & 0 & 0 & 1 & 0 & 0 & 1 \end{bmatrix}$$

The matrices $\underset{\sim}{K}_{21}$, $\underset{\sim}{K}_{22}$ are obtained similarly.

The operators $R_{ij}(I+R_{jj})^{-1}$ are said to represent the column subsystems corresponding to the system description (27). As shown in the next section, column subsystems play a key role in the decomposition of large-scale systems.

4.3 RESULTS ON WELL-POSEDNESS AND STABILITY

In this section, we show that once a large-scale interconnected system (LSIS) has been decomposed into an interconnection of strongly connected subsystems, the well-posedness and stability of the overall system can be ascertained by studying only the isolated subsystems. The advantages of such results are readily apparent, because the union of all isolated subsystems is in general simpler than the original system.

To begin with, we relate the properties of the system

1a $$d_i = x_i - R_{ii} z_i - \sum_{j=1}^{i-1} R_{ij} z_j$$

1b $$z_i = F_i d_i$$

$$\left.\right\} \quad i = 1,\ldots,k \qquad (S)$$

to those of the systems

2a $$d_i = x_i - R_{ii} z_i \qquad (S_i)$$

2b $$z_i = F_i d_i$$

3 Theorem Consider the system (S). Suppose that for all $i > j$, (i) the operator R_{ij} is causal, and (ii) for each finite T , there exists a finite constant k_{ijT} such that the incremental gain of $P_T R_{ij}$ is less than or equal to k_{ijT} . Under these conditions, the system (S) is well-posed if and only if each of the isolated subsystems (S_i) is well-posed.

Proof "if" Suppose each of the systems (S_i) is well-posed. We show first of all that, corresponding to each input set $x = (x_1,\ldots,x_k) \in L^n_{pe}$, there exist a unique $d \in L^n_{pe}$ and a unique $z \in L_{pe}^{n}$ such that (1) is satisfied. Since (1) is actually a collection of k equations, we prove this by induction. First of all, for $j = 1$, (1) is

4 $$z_1 = F_1 d_1$$

which is the same as the isolated subsystem (S_1); hence existence and uniqueness follows by the hypothesis that all subsystems (S_i) are well-posed. Now suppose that for $j = 1,\ldots,i-1$, (1) has exactly one solution for $d_1,\ldots,d_j$ and $z_1,\ldots,z_j$. For $j = i$, we have

5a $$d_i = x_i - R_{ii} z_i - \sum_{j=1}^{i-1} R_{ij} z_j = x_i' - R_{ii} z_i$$

5b $$z_i = F_i d_i$$

where

6 $$x_i' \triangleq x_i - \sum_{j=1}^{i-1} R_{ij} z_i$$

Now note that x_i' is uniquely determined, by the inductive hypothesis, and that $x_i' \in L_{pe}^{\nu_i}$. Since (S_i) is well-posed, (5) determines $d_i \in L_{pe}^{\nu_i}$ and $z_i \in L_{pe}^{\nu_i}$ uniquely. This shows that the inductive process can be continued. Hence the system (1) exhibits existence and uniqueness of solutions. To prove that d and z depend causally on x, we again proceed by induction. Suppose d_j, z_j, $j = 1,\ldots,i-1$ depend causally on $x_1,\ldots,x_{i-1}$. Then it is clear form (6) that x_i' depends causally on $x_1,\ldots,x_i$, because R_{ij} is causal whenever $i > j$. Next, since the system (S_i) is well-posed, d_i, z_i depend causally on x_i' and hence on $x_1,\ldots,x_i$. The global Lipschitz continuity of $P_T d$ and $P_T z$ as functions of $P_T u$ is proved in the same way. Hence the system (1) is well-posed.

"only if" Suppose that systems $(S_1),\ldots(S_{i-1})$ are well-posed but (S_i) is not. We show that the system (1) is also not well-posed.

To be specific, suppose that (S_i) is not well-posed because it violates condition (W1) of Definition (2.2.9), i.e., that for some specific input $x_{i0} \in L_{pe}^{\nu_i}$, (2) does not have a unique solution. If (S_i) violates either (W2) or (W3), the arguments below are easily modified. Let $x_j = 0$ for $j = 1,\ldots,$ i-1; by the "if" part of the proof above and the assumption that $(S_1),\ldots,(S_{i-1})$ are well-posed, we see that there exist unique $d_1,\ldots,d_{i-1}$ and $z_1,\ldots,z_{i-1}$ such that (1) is satisfied for $i = 1,\ldots,i-1$. Now let

7 $$x_i = x_{i0} + \sum_{j=1}^{i-1} R_{ij} z_j$$

With this input, the i-th equation in (1) becomes

8a $$d_i = x_{i0} + \sum_{j=1}^{i-1} R_{ij} z_j - R_{ii} z_i - \sum_{j=1}^{i-1} R_{ij} z_j = x_{i0} - R_{ii} z_i$$

8b $$z_i = F_i d_i$$

By assumption, (8) does not have a unique solution. Hence system (1) is not well-posed. □

Remarks (i) It is clear from the proof of Theorem (3) that, if we are only interested in the existence and uniqueness of solutions to (1), we can state the following result: "(S) exhibits existence and uniqueness of solutions corresponding to each $x \in L_{pe}^n$ if and only if each system (S_i) has exactly one solution corresponding to each $x_i \in L_{pe}^{n_i}$". In other words, we do not need to make any assumptions about R_{ij} if we are only interested in existence and uniqueness of solutions.

(ii) The hypothesis on R_{ij} in Theorem (3) imply that R_{ij} is a weakly Lipschitz operator whenever $i > j$ (see Definition (5.1.1)).

9 Theorem With respect to the system (S), suppose that

(i) For each input set $x \in L_{pe}^n$, (1) has a unique solution for d, z in L_{pe}^n.

(ii) $\gamma(R_{ij}) < \infty \quad \forall i > j$ (recall Definition (3.1.1)). Under these conditions, the system (S) is L_p-stable if and only if each of the isolated subsystems (S_i) is L_p-stable.

<u>Proof</u> "if" To avoid confusion, we use the superscript (I) to denote quantities associated with the isolated subsystems. Suppose each of the subsystems (S_i) is L_p-stable, and suppose k_i, b_i are finite constants such that

10 $$||d_i^{(I)}||_p, ||z_i^{(I)}||_p \leq k_i||x_i||_p + b_i$$

whenever x_i, $d_i^{(I)}$, $z_i^{(I)}$ satisfy (2). Now let $x \in L_p^n$ be an arbitrary input to the system (S). Then $d_1 = d_1^{(I)}$, $z_1 = z_1^{(I)}$. Hence from (10), we get

11 $$||d_1||_p, ||z_1||_p \leq k_1 ||x_1||_p + b_1$$

For $i = 2$, we have from (1) that

12 $$||d_2||_p, ||z_2||_p \leq k_2(||x_2 - R_{21} z_1^{(I)}||_p) + b_2$$
$$\leq k_2(||x_2||_p + k_1 ||x_1||_p + b_1) + b_2$$

The proof by induction is obvious.

"only if" Suppose the system (1) is L_p-stable. Then there exist finite constants k and b such that

13 $$||d||_p, ||z||_p \leq k ||x||_p + b$$

whenever x, d, z satisfy (1). In particular, let $x = 0$ in (1) and let $\bar{d}, \bar{z}$ denote the corresponding solutions of (1). Then (13) implies that $||\bar{d}||_p \leq b$, $||\bar{z}||_p \leq b$. Now let $x_i \in L_p^{\nu_i}$ be any input to the system (2), and let $\bar{x} = (0,0,0,\ldots,\bar{x}_i,0,\ldots,0) \in L_p^n$ be the "corresponding" input to the system (1), where

14 $$\bar{x}_i = x_i + \sum_{j=1}^{i-1} R_{ij} \bar{z}_j$$

By the hierachical arrangement of the subsystems in (1), we see that with the input x to the system (1), we still have $d_j = \bar{d}_j$, $z_j = \bar{z}_j$ for $j = 1,\ldots,i-1$. Moreover, by (13) we have

15 $$||d_i||_p \leq ||d||_p \leq k\ ||\bar{x}||_p + b = k\ ||\bar{x}_i||_p + b$$

$$\leq k\ ||x_i||_p + b_0$$

where

16 $$b_0 \triangleq b + k\ ||\sum_{j=1}^{i-1} R_{ij}\ \bar{z}_j||_p$$

Similarly we can show that

17 $$||z_i||_p \leq k\ ||x_i||_p + b_0$$

Now, with the input x_i to the system (2), $d_i^{(I)}$ and $z_i^{(I)}$ satisfy

18a $$d_i^{(I)} = x_i - R_{ii}\ z_i^{(I)}$$

18b $$z_i^{(I)} = F_i\ x_i$$

However, $d_i^{(I)} = d_i$ and $z_i^{(I)} = z_i$ because

19a $$d_i = \bar{x}_i - \sum_{j=1}^{i-1} R_{ij}\ z_j - R_{ii}\ z_i = x_i - R_{ii}\ z_i$$

19b $$z_i = F_i\ d_i$$

Hence from (15) and (16) we get

20 $$||d_i^{(I)}||_p\ ,\ ||z_i^{(I)}||_p \leq k\ ||x_i||_p + b$$

which shows that the system (S_i) is L_2-stable. □

Theorem (3) states that if each <u>interconnection</u> operator R_{ij} is weakly Lipschitz, then the LSIS (S) is well-posed if and only if each of the subsystems (S_i) is well-posed. Theorem

(9) shows that, if each <u>interconnection</u> operator R_{ij} has finite gain, and the overall system (S) (or, equivalently, each subsystem (S_i) exhibits existence and uniqueness of solutions, then the overall system (S) is L_p-stable if and only if each subsystem (S) is L_p-stable. In both cases, it is clear that one only needs to study the subsystems corresponding to the strongly connected components, in order to determine well-posedness and stability. Therefore, throughout the rest of this book, we assume without loss of generality that all LSIS's we study have strongly connected system digraphs. It is clear that, since the union of the isolated subsystems is no more complex than the original LSIS, the task of analyzing all the isolated subsystems individually is no more complex, and is in general less complex, than analyzing the original LSIS.

We now turn to the study of systems of the form

21 $$z_{ij} = \tilde{R}_{ij}\,(I+R_{jj})^{-1}\,(u_j - \sum_{k=1}^{j-1} K_{jk}\,z_{jk})$$

The results in this case are far neater than Theorems (3) and (9).

22 <u>Theorem</u> Consider a system described by (21). This system is well-posed if and only if each of the column subsystem operators defined by

23 $$\tilde{R}_{ij}(I+R_{jj})^{-1}$$

is well-posed; i.e., (i) $\tilde{R}_{ij}(I+R_{jj})^{-1}$ is a well-defined causal operator mapping $L_{pe}^{n_j}$ into $L_{pe}^{n_i n_j}$, (ii) the incremental gain of $P_T\,\tilde{R}_{ij}(I+R_{jj})^{-1}$ is finite for all T.

<u>Proof</u> "if" Suppose each of the column subsystem is well-posed. Then, in particular, we have that $\tilde{R}_{jj}(I+R_{jj})^{-1}$ is a well-posed operator for all j, and hence $K_{jj}\,\tilde{R}_{jj}(I+R_{jj})^{-1} = R_{jj}(I+R_{jj})^{-1}$ is a well-posed operator for all j. Since

24 $$I = (I+R_{jj})(I+R_{jj})^{-1} = (I+R_{jj})^{-1} + R_{jj}(I+R_{jj})^{-1}$$

we have

25 $$(I+R_{jj})^{-1} = I - R_{jj}(I+R_{jj})^{-1}$$

which shows that $(I+R_{jj})^{-1}$ is well-posed for all j. The rest of the proof essentially follows that of Theorem (3).

"only if" Parallel to that of Theorem (3). □

In comparing Theorem (3) and (22), we note that in Theorem (3), we are obliged to impose some hypotheses on the interconnection operators R_{ij}. In the system description (21), the role of the interconnection operators is played by K_{ij}, which can be represented by constant matrices , and therefore have a host of nice properties. Similar remarks apply to the stability results, proved next.

26 Theorem Consider a system described by (21), and suppose this system is well-posed. Then this system is L_p-stable if and only if each of the column subsystem operators $\tilde{R}_{ij}(I+R_{jj})^{-1}$ is L_p-stable for all i,j.

Proof Note that the assumption of well-posedness insures that $(I+R_{jj})^{-1}$ exists as a causal operator. The theorem is now obvious from Theorem (9). □

NOTES AND REFERENCES

Much of the background material on directed graphs follows the development of Baase [Baa. 1]. The results on decomposition into strongly connected components are due to Callier, Chan and Desoer [Cal. 1,2].

CHAPTER 5: WELL-POSEDNESS OF LARGE-SCALE INTERCONNECTED SYSTEMS

In this chapter, we state and prove some sufficient conditions for the well-posedness of large-scale interconnected systems. The conditions given here are structural in nature, in that they are graph-theoretic, and depend on classifying the various subsystem operators into two types. As such, these conditions are by no means necessary; however, they are very easy to verify. Moreover, if these sufficient conditions are not satisfied, it is easy to determine what "compensation" needs to be introduced in various parts of the system in order that these conditions be satisfied.

In this chapter, we treat discrete-time systems separately from continuous-time systems. The reason is that, even though the well-posedness criteria are essentially the same in both cases, the details of the proof are somewhat different. In contrast, we do not explicitly mention discrete time systems in Chapters 6-9, because the stability and instability results for discrete-time systems essentially parallel those for continuous-time systems.

5.1 WEAKLY LIPSCHITZ, SMOOTHING AND STRICTLY CAUSAL OPERATORS

The well-posedness results given here depend on the concepts of a weakly Lipschitz or smoothing operator, and a strictly causal operator. Therefore we begin by defining these concepts precisely.

1 <u>Definition</u> Let $p \in [1,\infty]$ and let n,m be given integers. We say that an operator $R: L_{pe}^n \to L_{pe}^m$ is <u>weakly Lipschitz</u> if it is causal, and if for every $T > 0$, there exists a finite constant K_T such that $\eta_p(P_T R) \leq k_T$, where $\eta_p(.)$ denotes the incremental gain of an operator (see Definition (3.1.1)). We say that an operator $R: L_{pe}^n \to L_{pe}^m$ is <u>smoothing</u> if it is weakly Lipschitz, and in addition, for every $T > 0$ and $\alpha > 0$, there exists a $\delta = \delta(\alpha,T)$ in $(0,T)$ such that

2 $$\eta_p[P_{t+\delta}(RP_{t+\delta} - RP_t)] \leq \alpha , \quad \forall t \in [0,T-\delta]$$

Remarks

1) Note that, because of causality, $\eta_p(P_tR) \leq \eta_p(P_TR)$ $\forall t \in [0,T]$. Hence we can also define a weakly Lipschitz operator as follows: an operator $R : L^n_{pe} \to L^m_{pe}$ is weakly Lipschitz if it is causal, and for every $T > 0$ there exists a finite constant k_T such that $\eta_p(P_tR) \leq k_T$ $\forall t \in [0,T]$.

2) If R is causal, it is easy to see that $[(RP_{t+\delta} - RP_t)x](s) = 0$ $\forall s \in [0,t]$. Hence $\{[P_{t+\delta}(RP_{t+\delta} - RP_t)]x\}(s) = 0$ whenever $s \notin [t,t+\delta]$. Thus, an operator R is smoothing if the dependence of $(Rx)(s)$, $s \in [t,t+\delta]$ on $x(s)$, $s \in [t,t+\delta]$ can be made to have arbitrarily small incremental gain by making δ sufficiently small (see Examples (9) and (21) below.

3 Example Suppose $\phi : R_+ \times R^n \to R^n$ satisfies the following condition: For every $T > 0$, there exists a constant k_T such that

4 $$||\phi(t,x) - \phi(t,y)|| \leq k_T \, ||x-y||, \; \forall x,y \in R^n, \; \forall t \in [0,T]$$

Then the memoryless operator $R_1 : L^n_{pe} \to L^n_{pe}$ defined by

5 $$(R_1x)(t) = \phi(t,x(t))$$

is weakly Lipschitz. To see this, observe that, whenever $x(.)$, $y(.) \in L^n_{pe}$, we have

6 $$||R_1x - R_1y||_{Tp} \leq k_T \, ||x-y||_{Tp}, \; \forall T$$

This shows that $\eta_p(P_TR_1) \leq k_T$, $\forall T$.

7 Example Let $\theta \geq 0$. Then the delay operator $R_2 : L^n_{pe} \to L^n_{pe}$ defined by

8 $$(R_2x)(t) = \begin{cases} 0, & t < \theta \\ x(t-\theta) & t \geq \theta \end{cases}$$

is weakly Lipschitz, because $\eta_p(P_TR) \leq 1$ $\forall T$.

9 <u>Example</u> Suppose $F : R_+ \times R_+ \to R^{m\times n}$ is continuous. Then the operator $R_3 : L^n_{pe} \to L^m_{pe}$ defined by

10 $$(R_3 x)(t) = \int_0^t F(t,\tau)\, x(\tau)\, d\tau$$

is smoothing. To see this, observe that

11 $$\{[P_{t+\delta}(R_3 P_{t+\delta} - R_3 P_t)]x\}(s) = \begin{cases} \int_t^s F(s,\tau)\, x(\tau)\, d\tau, & s \in [t,t+\delta] \\ 0 & \text{otherwise} \end{cases}$$

Hence, whenever $t \in [0, T-\delta]$, we have

12 $$||P_{t+\delta}(R_3 P_{t+\delta} - R_3 P_t)x||_p = \{ \int_t^{t+\delta} || \int_t^s F(s,\tau)\, x(\tau)\, d\tau||^p\, ds\}^{1/p}$$

Now, following the same reasoning as in the proof of Lemma (3.1.15), we get

13 $$||P_{t+\delta}(R_3 P_{t+\delta} - R_3 P_t)x||_p \le c_1^{1/p}\, c_\infty^{1/q}\, ||x||_p$$

where

14a $$c_1 = \sup_{\tau \in [t,t+\delta]} \int_\tau^{t+\delta} ||F(s,\tau)||\, ds$$

14b $$c_\infty = \sup_{s \in [t,t+\delta]} \int_t^s ||F(s,\tau)||\, d\tau$$

If we now define

15 $$M_T = \sup_{0 < s,\, \tau \le T} ||F(s,\tau)||$$

it is easy to see that $c_1 \le M_T \delta$, $c_\infty \le M_T \delta$. Substituting this in (13), and observing that since $P_{t+\delta}(R_3 P_{t+\delta} - R_3 P_t)$ is linear, its incremental gain is the same as its gain, gives

16 $$\eta_p[P_{t+\delta}(R_3P_{t+\delta}-R_3P_t)] \le M_T\delta \ , \ \forall t \in [0,T-\delta]$$

If we let $t = 0$ and $\delta = T$ in (16), we get

17 $$\eta_p(P_TR_3) \le T\ M_T$$

Now, (17) shows that R_3 is weakly Lipschitz, while (16) shows that R_3 satisfies (2). Hence R_3 is smoothing.

It is immediate from Definition (1) that if G and H are both weakly Lipschitz (smoothing), then so are $G \pm H$. Also, if G and H are both weakly Lipschitz, then so are GH and HG. Finally, we have

18 Lemma Suppose $G : L_{pe}^n \to L_{pe}^m$ is smoothing and that $H : L_{pe}^{\ell} \to L_{pe}^n$ is weakly Lipschitz. Then GH is smoothing.

Proof Clearly GH is weakly Lipschitz, since both G and H are weakly Lipschitz. So it only remains to show that GH satisfies (2). Accordingly, suppose $T > 0$ and let $\alpha > 0$ be specified. Since H is weakly Lipschitz, $\eta_p(P_TH) \triangleq \eta_T$ is finite, and also $\eta_p(P_tH) \le \eta_T$ $\forall t \in [0,T]$. Now pick $\delta \in (0,T)$ such that

19 $$\eta_p[P_{t+\delta}(GP_{t+\delta}-GP_t)] \le (\alpha/\eta_T), \ \forall t \in [0,T-\delta]$$

This is always possible because G is smoothing. Now, making repeated use of causality, we get

20 $$\begin{aligned}
&\eta_p[P_{t+\delta}(GHP_{t+\delta}-GHP_t)] \\
&= \eta_p[P_{t+\delta}(GP_{t+\delta}HP_{t+\delta}-GP_tHP_t)] \\
&= \eta_p[P_{t+\delta}(GP_{t+\delta}HP_{t+\delta}-GP_tHP_{t+\delta})] \\
&= \eta_p[\{P_{t+\delta}(GP_{t+\delta}-GP_t)\}.\ P_{t+\delta}HP_{t+\delta}] \\
&\le \eta_p[P_{t+\delta}(GP_{t+\delta}-GP_t)].\ \eta_p(P_{t+\delta}H) \\
&\le \frac{\alpha}{\eta_T}\cdot \eta_T = \alpha \ , \quad \forall t \in [0,T-\delta]
\end{aligned}$$

where in the third step we also use the facts that $P_{t+\delta} = P^2_{t+\delta}$, $P_t = P_t \cdot P_{t+\delta}$. Thus GH is smoothing. □

21 <u>Lemma</u> Let $G : L^n_{pe} \to L^m_{pe}$ be smoothing, and let H be a delay of the form (8). Then HG is smoothing.

<u>Proof</u> Clearly HG is weakly Lipschitz. Also,

22 $$P_{t+\delta}(HGP_{t+\delta} - HGP_t) = P_{t+\delta-\theta}(GP_{t+\delta} - GP_t).$$

Now, by causality,

23 $$\eta_p[P_{t+\delta-\theta}(GP_{t+\delta} - GP_t)] \leq \eta_p[P_{t+\delta}(GP_{t+\delta} - GP_t)],\ \forall t \in [0, T-\delta]$$

Since G is smoothing, the quantity on the right side of (23) can be made arbitrarily small by making δ small. Hence HG is smoothing. □

24 <u>Example</u> Suppose $F : R_+ \times R_+ \to R^{n\times n}$ is continuous, that ϕ satisfies (4), and that $p \in [1,\infty)$. Then $R_4 : L^n_{pe} \to L^n_{pe}$ defined by

25 $$(R_4 x)(t) = \int_0^t F(t,\tau)\ \phi(x(\tau))\ d\tau$$

is smoothing. This follows readily from Examples (3) and (9) and Lemma (18).

In studying discrete-time systems, the concept of a smoothing operator is naturally replaced by that of a strictly causal operator, defined next.

26 <u>Definition</u> An operator $G : S^n \to S^n$ is <u>causal</u> if $P_i G = P_i GP_i\ \forall i > 0$, and is <u>strictly causal</u> if $P_{i+1} GP_i = P_{i+1} G\ \forall i \geq 0$.

The above definition states that G is causal if, whenever u and v are two inputs in S^n such that for some $i \geq 0$ we have $P_i u = P_i v$, then $P_i Gu = P_i Gv$; G is strictly causal if $P_{i+1} Gu = P_{i+1} Gv$. It is clear that G is a causal operator if $(Gu)^{(i)}$ depends only on $u^{(0)}, u^{(1)}, \ldots, u^{(i)}$, whereas G is a strictly causal if $(Gu)^{(i)}$ depends only on $u^{(0)},\ldots,u^{(i-1)}$. In particular, if G is strictly causal, then $(Gu)^{(0)}$ is actually independent of u.

In the case of continuous-time systems, G is a smoothing opreator if $\eta_p[P_{t+\delta}(GP_{t+\delta}-GP_t)]$ can be made arbitrarily small by making δ arbitrarily small. However, in discrete-time systems, δ can only assume integer values, so that one cannot define a smoothing operator in this way. To get a suitable definition, we go back to the continuous-time case, and study the operator R_4 of (25). Let $u_1(.)$, $u_2(.)$ be two inputs such that $u_1(t) = u_2(t)$ for $t \in [0,T]$, but $u_1(T^+) \neq u_2(T^+)$. In this case, we still have $(R_4u_1)(T^+) = (R_4u_2)(T^+)$, and hence that name "smoothing operator". Correspondingly, in the discrete-time case, if $u_1^{(j)} = u_2^{(j)}$ for $j = 0,\ldots,i$ but $u_1^{(i+1)} \neq u_2^{(i+1)}$, we should still have $(Gu_1)^{(j)} = (Gu_2)^{(j)}$ for $j = 0,\ldots,i+1$. This is exactly the concept of strict causality.

Note that strict causality and its implications on feedback stability are discussed in [Sae. 1].

The following lemmas are easy to prove.

27 Lemma Suppose $G : S^n \to S^n$ is causal and $H : S^n \to S^n$ is strictly causal. Then HG is strictly causal.

28 Lemma Suppose $G : S^n \to S^n$ is strictly causal and let $D_i : S^n \to S^n$ be a delay of i units, $i \geq 0$. Then $D_i G$ is strictly causal.

29 Lemma Suppose $G : S^n \to S^n$ is causal, and D_i, $i \geq 1$, be a delay of i units. Then D_iG is strictly causal.

Note that Lemma (29) does not have continuous time counterpart.

To motivate the use of the preceding concepts in analyzing well-posedness, we present two lemmas. The proofs are omitted, as these lemmas are actually special cases of results presented subsequently (Lemma (30) is a special case of Lemma

(5.3.1) while Lemma (31) can be proved by adapting the proof of Theorem (5.4.1)).

30 Lemma Suppose $G : L^n_{pe} \to L^n_{pe}$ is smoothing. Then $(I+G)^{-1}$ exists and is weakly Lipschitz.

31 Lemma Suppose $G : S^n \to S^n$ is strictly causal. Then $(I+G)^{-1}$ exists (as an operator).

5.2 SINGLE-LOOP SYSTEMS

In this section, we state two simple sufficient conditions for the well-posedness of feedback systems.

1 Theorem Consider a single-loop feedback system described by

2a $$e_1 = u_1 - y_2$$

2b $$e_2 = u_2 + y_1$$

2c $$y_1 = G_1 e_1$$

2d $$y_2 = G_2 e_2$$

where $u_1, u_2, e_1, e_2, y_1, y_2$ all belong to L^n_{pe} for some fixed $p \in [1,\infty]$ and some positive integer n, and G_1, G_2 map L^n_{pe} into itself. Under these conditions, the system (2) is well-posed if both G_1 and G_2 are weakly Lipschitz, and at least one of them is smoothing.

3 Theorem Consider a single-loop feedback system described by (2), where $u_1, u_2, e_1, e_2, y_1, y_2$ all belong to S^n for some positive integer n, and G_1, G_2 map S^n into itself. Under these conditions, the system (2) is well-posed if both G_1 and G_2 are causal, and at least one of them is strictly causal.

The proofs of these theorems are omitted, as they are special cases of the results in the next section. The purpose of stating these theorems here is to show how the "large-scale" results are natural generalizations of the single-loop results.

Note that Theorem (1) is a better result than a similar one in [Des. 2, p. 48], while Theorem (3) has not previously appeared in this form.

5.3 CONTINUOUS-TIME SYSTEMS

In this section, we present the main result on the well-posedness of continuous-time large-scale interconnected systems. To do this, we require a preliminary

1 Lemma Consider a system described by the equations

2a $$d = H(u,z)$$

2b $$z = Gd$$

where $u \in L^n_{pe}$, $d \in L^{r_1}_{pe}$, $z \in L^{r_2}_{pe}$ for some fixed $p \in [1,\infty]$ and some positive integers n, r_1, r_2; $H : L^n_{pe} \times L^{r_2}_{pe} \to L^{r_1}_{pe}$ is weakly Lipschitz and $G : L^{r_1}_{pe} \to L^{r_2}_{pe}$ is smoothing. Then the system (2) is well-posed.

Proof We first establish that, corresponding to each $u \in L^n_{pe}$, (2) has a unique solution for (d,z) in $L^{r_1}_{pe} \times L^{r_2}_{pe}$. Now, (2) can be rewritten as

3 $$z = GH(u,z)$$

For fixed $u \in L^n_{pe}$, $H(u,.)$ is weakly Lipschitz and G is smoothing, so that by Lemma (4.2.18), the operator $z \mapsto GH(u,z)$ is smoothing. Let $T > 0$ be arbitrary and let $\alpha < 1$. Since $GH(u,.)$ is smoothing, we can find a $\delta \in (0,T)$ such that

4 $$\eta_p[P_{t+\delta}(GH(u,P_{t+\delta}\cdot) - GH(u,P_t\cdot))] \le \alpha \ , \ \forall t \in [0,T-\delta]$$

Now, by causality, (3) is equivalent to

5 $$P_t z = P_t GH(u,P_t z)$$

First of all, setting $t = 0$ in (4) shows that the map $P_\delta z \mapsto P_\delta GH(u,P_\delta z)$ is a contraction on the space $L_p^{r_2}[0,\delta]$.* Hence the equation

6 $$P_\delta z = P_\delta GH(u,P_\delta z)$$

has a unique solution for $P_\delta z$, so that the system equation (3) has a unique solution over $[0,\delta]$.

We now show that this solution can be extended to all of $[0,T]$. Suppose that a unique solution to (3) has been found over the interval $[0,t]$ where $t < T - \delta$; we show that this solution can be extended to $[0,t+\delta]$. From (3) and (5) we get

7 $$\begin{aligned} P_{t+\delta} z &= P_{t+\delta} GH(u,P_{t+\delta} z) \\ &= P_{t+\delta} GH(u,P_{t+\delta} z) - P_{t+\delta} GH(u,P_t z) + P_{t+\delta} GH(u,P_t z) \\ &= P_{t+\delta}[GH(u,P_{t+\delta} z) - GH(u,P_t z)] + P_{t+\delta} GH(u,P_t z) \end{aligned}$$

Now note that $P_{t+\delta} GH(u,P_t z)$ is a <u>known</u> quantity - call it v_t. With this definition, (7) is of the form

8 $$P_{t+\delta} z = P_{t+\delta}[GH(u,P_{t+\delta} z) - GH(u,P_t z)] + v_t$$

Now, because of (4), the mapping $P_{t+\delta} z \mapsto P_{t+\delta}[GH(u,P_{t+\delta} z) - GH(u,P_t z)]$ is a contraction on the space $L_p^{r_2}[0,t+\delta]$. So once again by applying the contraction mapping theorem, we see that

---

*Note that if $t = 0$, then $GH(u,0)$ is independent of z; hence $\eta_p[P_\delta(GH(u,P_\delta\cdot) - GH(u,0))] = \eta_p[P_\delta GH(u,P_\delta\cdot)] \le \alpha$.

(8) determines $P_{t+\delta}z$ uniquely. In this way, the unique solution can be extended to all of $[0,T]$. Since T is arbitrary, it follows that (3) has a unique solution for $z \in L_{pe}^{r2}$ corresponding to each $u \in L_{pe}^{n}$. Once z is uniquely determined, (2a) determines d uniquely.

Thus we have shown that the system (2) satisfies condition (W1) of Definition (1.3.9). Conditions (W2) and (W3) are easily verified, if one observes that $H(u,z)$ is causal in u and weakly Lipschitz in u. Thus all the conditions of Definition (2.2.9) are satisfied, and the system (2) is well-posed. □

<u>Remarks</u> By Lemma (5.1.18), G followed by a delay is still smoothing, while H followed by a delay is still weakly Lipschitz. Thus, the system (2) remains well-posed even if delays are introduced after G and H.

We now come to the main result of this section.

9 <u>Theorem</u> Consider a large-scale interconnected system described by

10 $$e_i = u_i - \sum_{j=1}^{m} S_{ij}\, e_j , \quad i = 1,\ldots,m$$

where $e_i, u_i \in L_{pe}^{n_i}$ for some fixed $p \in [1,\infty]$ and some positive integer n_i, and $S_{ij} : L_{pe}^{n_j} \to L_{pe}^{n_i}$. Suppose that the system digraph D corresponding to (10) is strongly connected, and that all operators S_{ij} are either weakly Lipschitz or smoothing. Delete all edges corresponding to smoothing operators from the system digraph D, and label the resulting digraph as the reduced digraph. Under these conditions, the system (10) is well-posed if the reduced digraph contains neither cycles nor self-loops.

<u>Remarks</u> The assumption that the system digraph D is strongly connected may appear to be overly restrictive, but in fact it results in no loss of generality. First of all, if D is <u>not</u> strongly connected, we can analyze the well-posedness of each strongly connected component of D individually (see

Theorem (4.3.22)). Alternatively, we can add edges corresponding to the zero operator, as required, in order to make D strongly connected; if this procedure is followed, it should be remembered that the zero operator is smoothing.

Proof Since the reduced digraph contains neither cycles nor self-loops, it can be incorporated into a directed tree; when this is done, all cotree edges (or links) correspond to smoothing operators.

Let the set of edges in the cotree be denoted by E . Let us label the input vertices of the edges in E as $d_1,\ldots,d_k$, and the output vertices of the edges in E as $z_1,\ldots,z_k$. By the input vertex of an edge, we mean the vertex from which the edge leaves, and by the output vertex of an edge, we mean the vertex to which the edge is directed. Now, some of the d_i's may be the same as some of the z_i's; also, if an edge in E is a self-loop, then we will have $d_i = z_i$ for some i . All this does not matter.

Let us now go back to the digraph D , which is the union of the directed tree containing weakly Lipschitz or smoothing operators, and a cotree containing only smoothing operators. Let us consider a typical cotree edge η_i in E , as shown in Figure 5.1. Suppose we break the edge η_i into three edges as shown in Figure 5.2, by introducing two new vertices d_i' and z_i'.

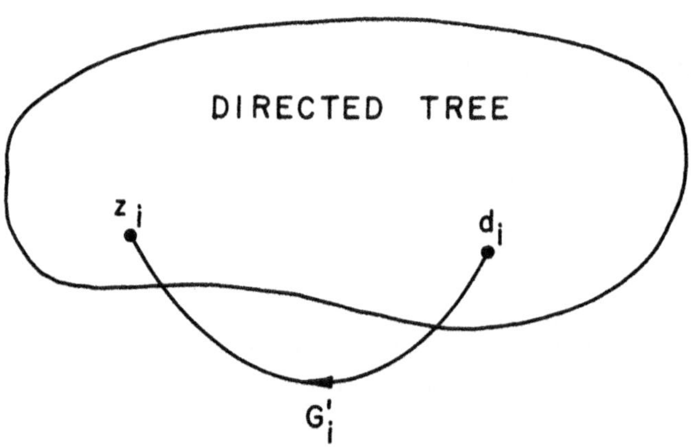

FIGURE 5.1

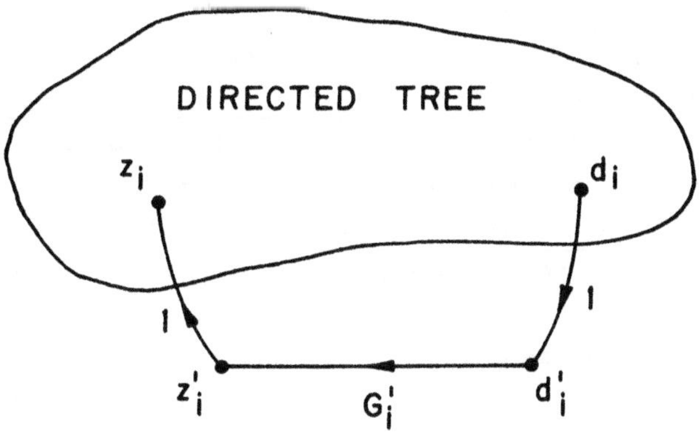

FIGURE 5.2

Since η_i is a cotree edge, the addition of η_i to the directed tree creates a unique cycle. Hence there is a path from z_i to d_i consisting only of edges from the directed tree. With reference to Figure 5.2, this means that there is a path from z_i' to d_i' . This is true of all edges in E . Moreover, it is clear that the path from z_i' to d_i' does not pass through d_j' for $j \neq i$. Thus we can express the quantity d_i' in terms of the vector $z' = (z_1', \ldots, z_k')$ and the input vector $u = (u_1, \ldots, u_m)$. Let this relationship be denoted

11a $$d_i' = F_i(u, z') \ , \quad i = 1, \ldots, k$$

Now, the operator F_i consists of sums and compositions of weakly Lipschitz operators, and is thus a weakly Lipschitz operator. From the digraph, we also have

11b $$z_i' = G_i'(d_i') \ , \quad i = 1, \ldots, k$$

where G_i' is a smoothing operator. Looking at equations (11), we can see that we have succeeded in rewriting the system equations in the form (2). Hence by Lemma (1), the system (10) is well-posed. □

Remarks Basically, the condition that the reduced digraph (obtained by removing all edges corresponding to smoothing operators) is acyclic insures that all cycles contain at least one edge corresponding to a smoothing operator. The method of proof makes it clear that if the hypotheses of Theorem (9) are satisfied, then one can find two sets of variables d_i' and z_i' such that the system equations can be rearranged in the form (11). It is also clear that, if a particular system (10) satisfies the hypotheses of Theorem (9) and is thus well-posed, and if we perturb each of the operators G_i , H_{ij} , the system continues to be well-posed so long as each smoothing operator is perturbed into another smoothing operator, and each weakly Lipschitz operator is perturbed into another weakly Lipschitz or smoothing operator.

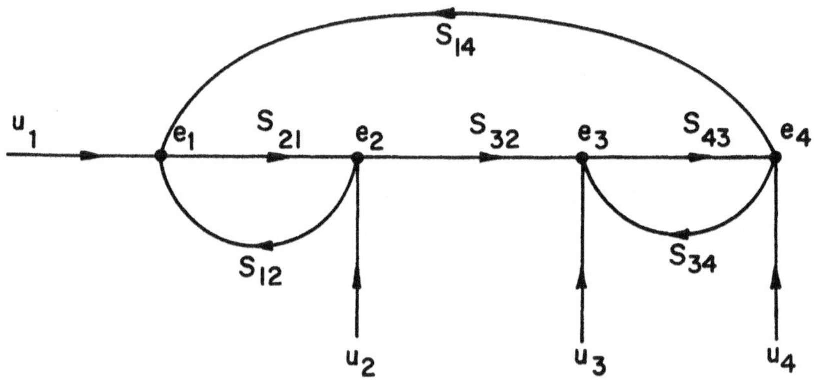

FIGURE 5.3

12 Example Consider the system of Figure 5.3, and suppose that S_{12} thru S_{43} are all weakly Lipschitz operators; the question is, which of these operators need to be smoothing in order for Theorem (9) to apply. Clearly, there are many possibilities. For instance, if S_{21} and S_{43} are smoothing, then we select d_1' and d_2' as shown in Figure 5.4, and the system equations become

13a $$d_1' = u_1+S_{12}(z_1'+u_2) + S_{14}(z_2'+u_4)$$

13b $$d_2' = u_3+S_{32}(z_1'+u_2) + S_{34}(z_2'+u_4)$$

13c $$z_1' = S_{21}\ d_1'$$

13d $$z_2' = S_{43}\ d_2'$$

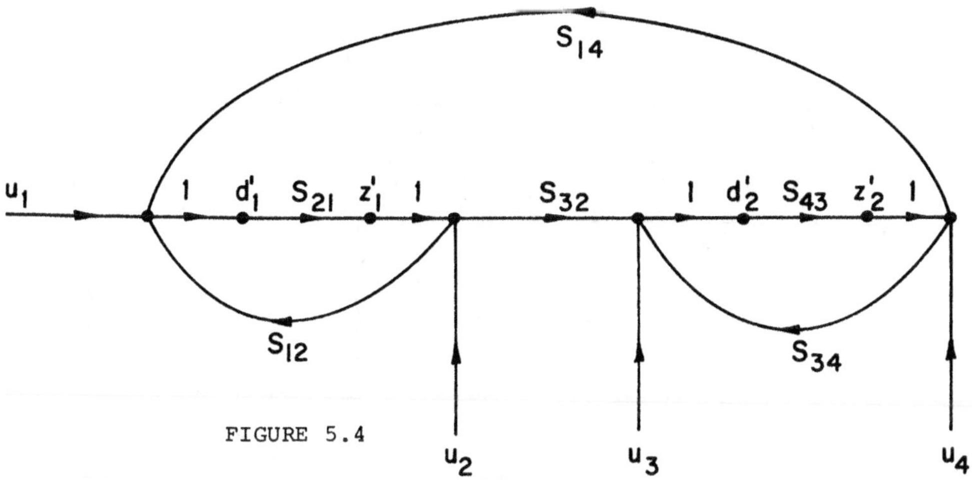

FIGURE 5.4

On the other hand, suppose S_{12}, S_{34} and S_{14} are smoothing. Then d_1' , d_2' , d_3' , z_1' , z_2' , z_3' would be chosen as in Figure 5.5, and the system equations would become

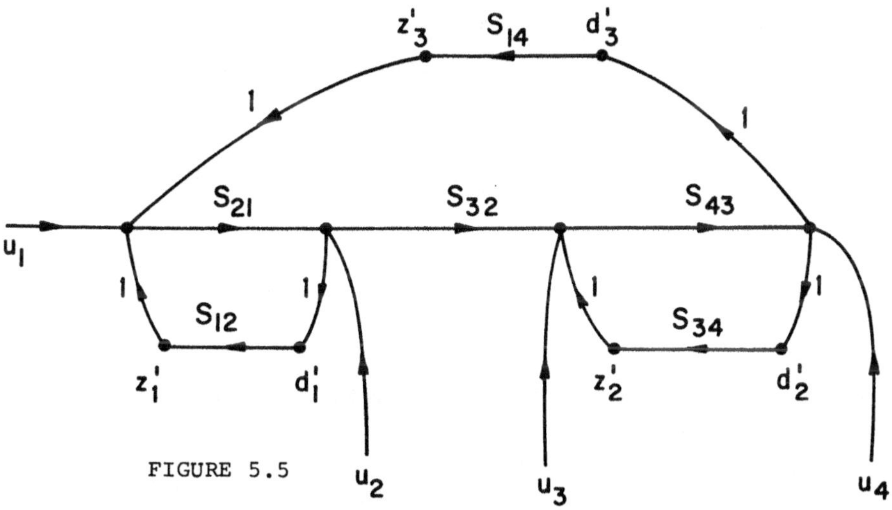

FIGURE 5.5

14a $$d_1' = u_2 + S_{21}(u_1+z_1'+z_3')$$

14b $$d_2' = u_4 + S_{13}(u_3+z_2'+S_{32}(u_2+S_{21}(u_1+z_1'+z_3')))$$

14c $$d_3' = d_2'$$

14d $$z_1' = S_{12}d_1' \text{ , } z_2' = S_{34}d_2' \text{ , } z_3' = S_{14}d_3'$$

Once again, the system equations are of the form (11).

15 Example In this example, we study a system with self-loops. Let the system under consideration be as in Figure 5.6 . It is easy to verify that if S_{11}, S_{21}, and S_{22} are smoothing

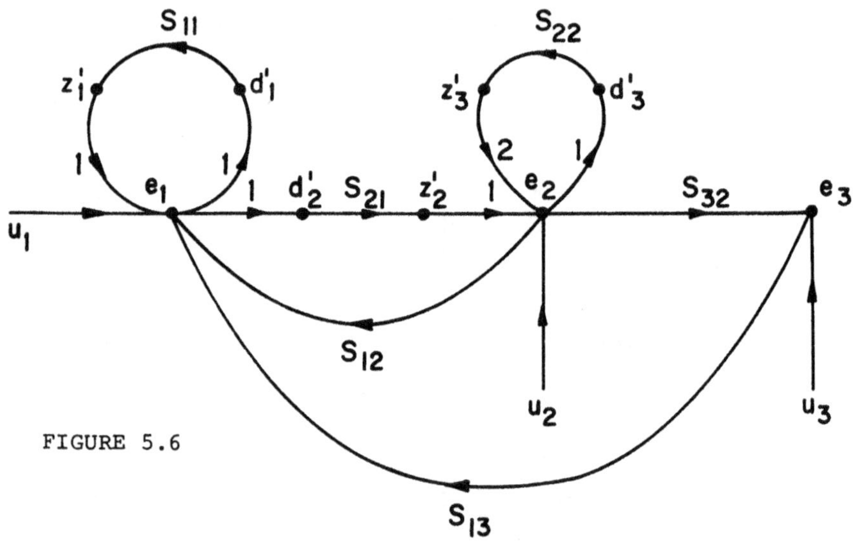

FIGURE 5.6

operators and the rest are weakly Lipschitz operators, then the reduced digraph is acyclic and does not contain any self-loops. Choosing d_1' thru z_3' as indicated in Figure 5.6, we get the system equations

16a $$d_1' = u_1 + S_{12}(u_2+z_2'+z_3')+z_1' + S_{13}(u_3+S_{32}(z_2'+u_2+z_3'))$$

16b $$d_2' = d_1'$$

16c $$d_3' = z_2' + u_2$$

16d $$z_1' = S_{11}d_1' , \quad z_2' = S_{21}d_2' , \quad z_3' = S_{22}d_3'$$

which are of the form (11).

It is clear from the above examples that, in order to satisfy the hypotheses of Theorem (10), all self-interaction operators S_{ii} must be smoothing.

Now we turn to systems described by

17a $$e_i = u_i - \sum_{j=1}^{m} H_{ij}\, y_j \qquad i = 1,\dots,m$$

17b $$y_i = G_i\, e_i$$

where $u_i, e_i, y_i \in L_{pe}^{n_i}$ for some fixed $p \in [1,\infty]$ and for some positive integer n_i, $G_i : L_{pe}^{n_i} \to L_{pe}^{n_i}$, and $H_{ij} : L_{pe}^{n_j} \to L_{pe}^{n_i}$. As shown in Section 2.2, the system (17) can be equivalently described by

18a $$e_i = u_i - \sum_{j=1}^{m} H_{ij}\, e_{j+m} \qquad \Bigg\} \; i = 1,\ldots,m$$

18b $$e_{i+m} = G_i\, e_i$$

Suppose now that all operators G_i, H_{ij} are weakly Lipschitz. Then the system (18) is well-posed in the sense of Definition (2.2.19) if and only if the system (17) is well-posed in the sense of Definition (2.2.9). (See (2.2.31) <u>et seq</u>.). Hence we can apply Theorem (1) to study the well-posedness of systems described in the form (17).

5.4 DISCRETE-TIME SYSTEMS

The graph-theoretic condition presented below for the well-posedness of discrete-time systems is the same as that for continuous-time systems. But the method of proof is quite different.

1 <u>Theorem</u> Consider a discrete-time system described by

2 $$e_i = u_i - \sum_{j=1}^{m} S_{ij}\, e_j\,, \quad i = 1,\ldots,m$$

where e_i, u_i belong to S^{n_i} for some positive integer n_i, and $S_{ij} : S^{n_j} \to S^{n_i}$. Suppose that for all i, j, S_{ij} is causal, that $P_k S_{ij}$ has finite incremental gain for all integers k, and that some of these operators are strictly causal. Associated with the system (2), construct a reduced digraph as in Theorem (5.3.9). Under these conditions, the system (2) is well-posed if the reduced digraph, obtained by removing all edges corresponding to strictly causal operators from the system digraph, does not contain either cycles or self-loops.

<u>Proof</u> Just as in the proof of Theorem (5.3.9), if the hypotheses of the theorem are satisfied, then we can find two sets of variables $\{d_1',\ldots,d_k'\}$ and $\{z_1',\ldots,z_k'\}$ such that the system equations can be rewritten into the form

3a $$d' = H(u,z')$$

3b $$z' = G\,d'$$

where H is causal and G is strictly causal, and both P_kG, P_kH have finite incremental gains for all integers k. So the theorem is proved if we show that a system of the form (3) is well-posed.

First of all, we prove that, corresponding to any $u \in S^n$, (3) has a unique solution for d' and z'. Now, since G is strictly causal, $(Gd')^{(0)}$ is independent of d', and hence $z'^{(0)}$ is known from (3b). Knowing $z'^{(0)}$ and $u^{(0)}$, we can find $d'^{(0)}$ from (3a). Then from (3b) we can find $z'^{(1)}$, and so on. Thus condition (W1) of Definition (2.2.20) is satisfied. Conditions (W2) - (W3) are easy to verify. Hence the system (2) is well-posed. □

The condition that the reduced digraph be acyclic implies that each cycle contains at least one edge corresponding to a strictly causal operator, i.e. that there are no "delay free loops". It is well-known in the theory of digital filters that if a filter has no delay-free loops, then it can be implemented in a recursive fashion. This is essentially what we do in the proof of Theorem (1).

NOTES AND REFERENCES

The contents of this Chapter closely follow [Vid. 9]. Well-posedness of feedback systems is studied by Willems [Wil. 2], while [Des. 2] contains some results on the existence and uniqueness of solutions in the case of feedback systems. State-space analogs of the results given here can be found in [Moy. 4].

CHAPTER 6: SMALL-GAIN TYPE CRITERIA FOR L_p-STABILITY

In this Chapter, we state and prove several sufficient conditions of the "small-gain" type for a large-scale interconnected system to be L_p-stable. The results presented here can be placed into two categories: (i) conditions involving a test matrix, and (ii) conditions involving a minimal essential set decomposition of the system digraph. It is interesting to note that both types of results reduce, in the single-loop case, to the standard small gain theorems of Section 6.1.

The results of this chapter can be used to establish L_p-stability for any value of p in $[1,\infty]$. In contrast, the results of Chapter 7 require the input space to have an inner product structure. In this respect, the results of this cahpter have a wider scope of application than those in Chapter 7. On the other hand, the present results, in most cases, are of "weak interconnection" or "worst-case" type, and can lead to overly conservative stability criteria. This point is elaborated in the sequel.

6.1 SINGLE-LOOP SYSTEMS

In this section, we present the standard small-gain theorems for single-loop systems. The proofs are omitted, as the theorems here are special cases of subsequent results.

1 Theorem Consider a single-loop feedback system described by the equations

2a $$e_1 = u_1 - y_2$$

2b $$e_2 = u_2 + y_1$$

2c $$y_1 = G_1 e_1$$

2d $$y_2 = G_2 e_2$$

where $u_1, u_2, e_1, e_2, y_1, y_2 \in L_{pe}^{\nu}$ for some fixed $p \in [1,\infty]$ and some positive integer ν, and G_1, G_2 map L_{pe}^{ν} into itself. Let $\gamma_p(G_1)$ and $\gamma_p(G_2)$ denote the gains of the operators G_1 and G_2, as defined in Definition (3.1.1). Under these conditions, the system (2) is L_p-stable if

3 $$\gamma_p(G_1)\ \gamma_p(G_2) < 1$$

4 Theorem Suppose that either $u_2 = 0$ or that G_2 is linear. Under these conditions, the system (2) is L_p-stable if $\gamma_p(G_1)$ and $\gamma_p(G_2)$ are finite, and further,

5 $$\gamma_p(G_2G_1) < 1$$

Remarks 1) Both conditions (3) and (5) are usually referred to as "small-gain" type conditions, as they require either the gain of an operator or a product gains (of operators) to be less than one. Note that (5) is a less restrictive condition than (3), since $\gamma_p(G_2G_1) \leq \gamma_p(G_2)\ \gamma_p(G_1)$. Hence, Theorem (4) gives a weaker sufficient condition for L_p-stability than Theorem (1), in cases where either $u_2 = 0$ or G_2 in linear.

2) Both the conditions (3) and (5) are "sign-insensitive." In other words, if (3) holds for a particular pair of operators G_1, G_2, then (3) is still satisfied if G_1 is replaced by $-G_1$ and/or G_2 is replaced by $-G_2$. More generally, if (3) holds for a particular pair of operators G_1, G_2, then (3) is still satisfied whenever G_1 is replaced by $\alpha_1 G_1$, G_2 is replaced by $\alpha_2 G_2$, and $\alpha_1, \alpha_2 \in [-1,1]$. These remarks follow from the obvious identity

6 $$\gamma_p(\alpha G) = |\alpha|\ \gamma_p(G)\ ,\ \forall\alpha$$

3) Theorem (4) holds if the subscripts 1 and 2 are interchanged throughout.

The following results use what is known as a "Loop transformation."

7 Theorem Consider the system (2), and let $K : L_{pe}^{\nu} \to L_{pe}^{\nu}$ be a linear operator such that $\gamma_p(K) < \infty$, and $(I+KG_1)$ has an inverse $(I+KG_1)^{-1} : L_{pe}^{\nu} \to L_{pe}^{\nu}$. Under these conditions, the

system (2) is L_p-stable if

8 $$\gamma_p(G_2-K) \cdot \gamma_p(G_1(I+KG_1)^{-1}) < 1$$

9 Theorem Consider the system (2) and let K be as in Theorem (7). Suppose that either $u_2 = 0$ or G_2 is linear. Under these conditions, the system (2) is L_p-stable if

10 $$\gamma_p[(G_2-K)\ G_1(I+KG_1)^{-1}] < 1 .$$

To apply any of the above theorems to a specific example, one has to be able to calculate the gain of a given operator. This task is facilitated by the results of Section 3.1.

6.2 CRITERIA BASED ON A TEST MATRIX

In this section, we study a large-scale interconnected system described by

1a $$e_i = u_i - \sum_{j=1}^{m} H_{ij}\, y_j$$

1b $$y_i = G_i\, e_i$$

$$\left.\right\} \quad i = 1,\ldots,m$$

where $u_i, e_i, y_i \in L_{pe}^{n_i}$ for some fixed $p \in [1,\infty]$ and some positive integer n_i, $G_i : L_{pe}^{n_i} \to L_{pe}^{n_i}$, and $H_{ij} : L_{pe}^{n_j} \to L_{pe}^{n_i}$. The stability criteria presented here all involve checking a so-called "test matrix" to verify whether all of its leading principal minors are positive.

6.2.1. RESULTS ON NONNEGATIVE MATRICES

We first present a few preliminary results on non-negative matrices.

2 Fact Given $A \in R^{n\times n}$, let $sp(A)$ denote the spectrum of A (i.e., the set of eigenvalues of A), and define the spectral radius $\rho(A)$ of A by

3 $$\rho(A) = \max_{\lambda \in Sp(A)} |\lambda|$$

Suppose $A \in R^{n\times n}$ has all nonnegative elements. Then $\rho(A) \in Sp(A)$.

<u>Proof</u> See [Gan. 1, p. 66].

4 <u>Fact</u> Let $B \in R^{n\times n}$ have all nonpositive off-diagonal elements (i.e. $b_{ij} \leq 0$ whenever $i \neq j$). Under these conditions, the following statements are all equivalent:

(i) All eigenvalues of B have positive real parts.

(ii) The leading principal minors of B are all positive.

(iii) There exist positive constants $d_1,\ldots,d_n$ such that $D\,B$ is strictly row-dominant, where $D = \text{Diag}\,\{d_1,\ldots,d_n\}$. In other words, there exist positive constants $d_1,\ldots,d_n$ such that

5 $$d_j\, b_{jj} > \sum_{\substack{i=1\\ i\neq j}}^{n} |d_i\, b_{ij}| = -\sum_{\substack{i=1\\ i\neq j}}^{n} d_i\, b_{ij}\,, \quad \forall j$$

One can also express (5) equivalently as

6 $$\sum_{i=1}^{n} d_i\, b_{ij} > 0\,, \quad \forall j$$

(iv) There exist positive constants $c_1,\ldots,c_n$ such that $B\,C$ is strictly column-dominant, where $C = \text{Diag}\,\{c_1,\ldots,c_n\}$. In other words, there exist positive constants $c_1,\ldots,c_n$ such that

7 $$\sum_{j=1}^{n} b_{ij}\, c_j > 0\,, \quad \forall i$$

In effect, (ii)-(iv) are all equivalent necessary and sufficient conditions for (i). The proof of Fact (4), plus several other equivalent necessary and sufficient conditions for (i), can be found in [Fie. 1] or more accessibly in [Sil. 1, p. 396].

8 **Lemma** Suppose $A \in R^{n\times n}$ has all nonnegative entries. Then $\rho(A) < 1$ if and only if the leading principal minors of $I_n - A$ are all positive.

Proof "if" By Fact (2), $1-\rho(A)$ is an eigenvalue of $I_n - A$. Hence, if the leading principal minors of $I_n - A$ are all positive, it follows from Fact (4) that $1-\rho(A) > 0$, i.e. $\rho(A) < 1$.

"only if" Suppose $\rho(A) < 1$. Then $|\text{Re}\ \lambda| < 1$ whenever $\lambda \in Sp(A)$, by the definition of the spectral radius. Therefore $1-\text{Re}\ \lambda > 0 \quad \forall\lambda \in Sp(A)$, so that all eigenvalues of $I_n - A$ have positive real parts. Hence, by Fact (4), it follows that all the leading principal minors of $I_n - A$ are positive.□

9 **Lemma** Suppose $A \in R^{n\times n}$ has all nonnegative elements. Then $(I_n - A)^{-1}$ has all nonnegative elements if the leading principal minors of $I_n - A$ are all positive.

Proof If the leading principal minors of $I_n - A$ are all positive, then $\rho(A) < 1$ by Lemma (8). Hence $I_n - A$ has an inverse, which is given by

10 $$(I_n - A)^{-1} = \sum_{i=0}^{\infty} (A)^i$$

which shows that $(I_n - A)^{-1}$ has all nonnegative elements. □

11 **Fact** Suppose $B \in R^{n\times n}$ has all nonpositive off-diagonal elements. Then there exists a diagonal matrix $\Lambda = \text{Diag}\ \{\lambda_1, \ldots, \lambda_n\}$ with $\lambda_i > 0\ \forall i$ such that $B'\Lambda + \Lambda B$ is positive definite, if and only if the leading principal minors of B are all positive.

Proof See [Ara. 1].

6.2.2 BASIC "TEST-MATRIX" TYPE CRITERIA

With the aid of these results, we can derive several stability criteria for the system (1) .

12 *Theorem* Suppose all operators G_i have finite gains for all i, and define the test matrix $Q_1 \in R^{m\times m}$ by

13 $$q_{1ij} = \gamma_p(H_{ij}\,G_j)$$

Assume that q_{1ij} is finite for all i,j. Then the system (1) is L_p-stable if all the leading principal minors of $I_m - Q_1$ are positive.

Remarks By Lemma (8), the hypotheses on Q_1 are equivalent to requiring that $\rho(Q_1) < 1$.

Proof By the definition of $\gamma_p(H_{ij}\,G_j)$, there exist finite constants b_{ij} such that

14 $$||H_{ij}G_j x||_{Tp} \le \gamma_p(H_{ij}G_j)\,||x||_{Tp} + b_{ij}, \quad \forall T \ge 0,\ \forall x \in L_{pe}^{n_j}$$

Substituting from (1b) into (1a), we can recast the system equations in the form

15 $$e_i = u_i - \sum_{j=1}^{m} H_{ij}\,G_j\,e_j\,, \quad i = 1,\ldots,m$$

Taking norms in (15) and applying (14), we get

16 $$\begin{aligned} ||e_i||_{Tp} &\le ||u_i||_{Tp} + \sum_{j=1}^{m} ||H_{ij}G_j e_j||_{Tp} \\ &\le d_i + ||u_i||_{Tp} + \sum_{j=1}^{m} \gamma_p(H_{ij}G_j)\,||e_j||_{Tp} \end{aligned}$$

where

17 $$d_i = \sum_{j=1}^{m} b_{ij}$$

Suppose now that $u_i \in L_p^{n_i}$ $\forall i$; we first show that $e_i, y_i \in L_p^{n_i}$ $\forall i$. Since $u_i \in L_p^{n_i}$, we can replace $||u_i||_{Tp}$ by its upper bound $||u_i||_p$ in (16). This gives

18 $$\|e_i\|_{Tp} \le d_i + \|u_i\|_p + \sum_{j=1}^{m} \gamma_p(H_{ij}\, G_j)\, \|e_j\|_{Tp}$$
$$= d_i + \|u_i\|_p + \sum_{j=1}^{m} q_{1ij}\, \|e_j\|_{Tp}$$

19 $$\sum_{j=1}^{m} (\delta_{ij}-q_{1ij})\, \|e_j\|_{Tp} \le d_i + \|u_i\|_p$$

where δ_{ij} denotes the Kronecker delta. Let us define vectors $v_T \in R^m$ and $w \in R^m$ by

20 $$v_T = [\|e_1\|_T \;\ldots\; \|e_m\|_T]'$$

21 $$w = [d_i + \|u_1\| \;\ldots\; d_m + \|u_m\|]'$$

Then (19) is equivalent to

22 $$[(I_m-Q_1)\, v_T]_i \le (w)_i$$

Since the leading principal minors of I_m-Q_1 are all positive, we have by Lemma (9) that $(I_m-Q_1)^{-1}$ contains all nonnegative elements. Hence we can multiply both sides of (22) by $(I_m-Q_1)^{-1}$; this gives

23 $$(v_T)_i = \|e_i\|_{Tp} \le [(I_m-Q_1)^{-1}\, w]_i$$

Since the right side of (23) is independent of T, it follows from Lemma (2.1.12) that $e_i \in L_p^{n_i}\ \forall i$. Since $\gamma_p(G_i) < \infty\ \forall i$, it follows that $y_i \in L_p^{n_i}\ \forall i$. Finally, it is clear from (23) and the finiteness of $\gamma_p(G_i)\ \forall i$ that there exist finite constants k, b such that

24a $$\|e\|_p \le k\, \|u\|_p + b$$

24b $$\|y\|_p \le k\, \|u\|_p + b$$

where $e = (e_1,\ldots,e_m)$ and u,y are defined similarly. Hence the system (1) is L_p-stable.□

The above proof makes it clear that, in applying Theorem (12), one can take q_{1ij} to be any <u>upper bound</u> for $\gamma_p(H_{ij}\, G_j)$, if the calculation of $\gamma_p(H_{ij}\, G_j)$ is too cumbersome.

In the single-loop case, Theorem (12) reduces to Theorem (6.1.1). To see this, apply Theorem (12) to the system (6.1.2), with $n = 2$, and

25 $$H_{11} = H_{22} = 0 \; ; \; H_{12} = -H_{21} = I$$

where I denotes the identity operator on L_{pe}^{ν}. Then we get

26 $$\gamma(H_{11}G_1) = \gamma(H_{22}G_2) = 0 \; ; \; \gamma(H_{12}G_2) = \gamma(G_2) \; ;$$
$$\gamma(H_{21}G_1) = \gamma(G_1)$$

27 $$Q_1 = \begin{bmatrix} 0 & \gamma(G_2) \\ \gamma(G_1) & 0 \end{bmatrix}$$

Thus the condition that the leading principal minors of $I_2 - Q_1$ are positive reduces to

28 $$\gamma(G_2) \cdot \gamma(G_1) < 1$$

which is the same as (6.1.3).

It is possible to "aggregate" the system equations (1) into the form

29a $$e = u - Hy$$

29b $$y = Ge$$

where

30 $$e = (e_1, \ldots, e_m)$$

31 $$G = \text{Diag}\,\{G_1, \ldots, G_m\}$$

32 $$H = \begin{bmatrix} H_{11} & \cdots & H_{1m} \\ \vdots & & \vdots \\ H_{m1} & \cdots & H_{mm} \end{bmatrix}$$

and u, y are defined analogously to (30). Applying Theorem (6.1.4) to the aggregated "single-loop" system description (29), we get

33 $$\gamma_p(HG) < 1$$

as a sufficient condition for the L_p-stability of the system (1).

It is interesting to compare the stability criterion (33) with that of Theorem (12). First of all, observe that it is possible to define the norm on the input space $L_p^n = \prod_{i=1}^{m} L_p^{n_i}$ in an infinite number of ways. For different choices of the norm on L_p^n, the number $\gamma_p(HG)$ will be different. Now, in order for the system (1) to be stable, it is sufficient for $\gamma_p(HG)$ to be less than one for <u>some</u> choice of the norm on L_p^n. Thus, (33) represents a very flexible stability criterion. If one simply defines the norm on L_p^n in the usual way (Definition 2.1.13), then the criterion (33) neither contains nor is contained by that of Theorem (12). On the other hand, if one takes full advantage of the flexibility offerred by (33), then (33) can be shown to contain Theorem (12) as a special case. To illustrate this point, we first give a

34 <u>Definition</u> Given positive constants $d_1, \ldots, d_m$, the norm $||\cdot||_{dp}$ on $L_p^n = \prod_{i=1}^{m} L_p^{n_i}$ is defined by

35 $$||x||_{dp} = \sum_{i=1}^{m} d_i \, ||x_i||_p \, , \quad \forall x = [x_1 \ldots x_m]' \in L_p^n$$

Given any $F : L_{pe}^n \to L_{pe}^n$, its gain $\gamma_{dp}(F)$ is defined by

36 $$\gamma_{dp}(F) = \inf \{k: \exists\, b < \infty \text{ such that } ||(Fx)_T||_{dp} \leq k\, ||x_T||_{dp}+b,\ \forall T \geq 0,\ \forall x \in L^n_{pe}\}$$

In other words, $||\cdot||_{dp}$ is just a particular type of norm on L^n_p, given by (35), and $\gamma_{dp}(F)$ is the corresponding gain of the operator F, calculated using $||\cdot||_{dp}$ as the norm on L^n_p. The relationship between (33) and Theorem (12) is brought out in the next

37 Lemma Given the system (1), define the matrix Q_1 as in (13). Under these conditions, if the leading principal minors of I_m-Q_1 are all positive, then there exist positive constants $d_1,\ldots,d_m$ such that $\gamma_{dp}(HG) < 1$.

Remarks Lemma (37) states that, if the stability of the system (1) can be established by Theorem (12), then it can also be established by applying the criterion (33) together with a suitably selected norm on L^n_p. In this sense, and in this sense only, Theorem (12) is a special case of the "small gain" criterion (33). However, it is clearly more easy to apply Theorem (12) than it is to calculate $\gamma_{dp}(HG)$ for several choices of $d_1,\ldots,d_m$ and seeing if any of these is less than one.

Proof Note that all elements of Q_1 are nonnegative, so that all the off-diagonal elements of I_m-Q_1 are nonpositive. Hence, if all the leading principal minors of I_m-Q_1 are positive, then there exist constants $d_1,\ldots,d_m$ such that

38 $$\sum_{i=1}^{m} d_i\ (\delta_{ij}-q_{1ij}) > 0 \quad \forall j$$

or, equivalently.

39 $$\sum_{i=1}^{m} d_i\ q_{1ij} < d_j \quad \forall j$$

Now pick a number $\alpha < 1$ such that

40 $$\sum_{i=1}^{m} d_i\ q_{1ij} \leq \alpha d_j \quad \forall j$$

This is clearly possible, in view of (39). We now claim that $\gamma_{dp}(HG) \leq \alpha < 1$. To show this, let $x = [x_1 \dots x_m]' \in L^n_{pe}$. Then

41 $$\begin{aligned} \|(HGx)_T\|_{dp} &= \sum_{i=1}^{m} d_i \, \|(HGx)_i\|_{Tp} && \text{(from (35))} \\ &= \sum_{i=1}^{m} d_i \, \|(\sum_{j=1}^{m} H_{ij} \, G_j x_j)\|_{Tp} && \text{(from(31)and(32))} \\ &\leq \sum_{i=1}^{m} d_i \, (\sum_{j=1}^{m} \gamma_p(H_{ij} G_j) \, \|x_j\|_{Tp} + b_{ij}) && \text{(from (14))} \\ &= \sum_{i=1}^{m} \sum_{j=1}^{m} d_i \, q_{1ij} \, \|x_j\|_{Tp} + b \\ &\leq \alpha \sum_{j=1}^{m} d_j \, \|x_j\|_{Tp} + b && \text{(from (40))} \\ &= \alpha \, \|x_T\|_{dp} + b \end{aligned}$$

where

42 $$b = \sum_{i=1}^{m} \sum_{j=1}^{m} d_i \, b_{ij}$$

The extreme inequalities in (41) show that $\gamma_{dp}(HG) \leq \alpha < 1$, which proves the lemma. □

43 <u>Theorem</u> Consider the system (1) , and define the test matrix $Q_2 \in R^{m \times m}$ by

44 $$q_{2ij} = \gamma_p(H_{ij}) \, \gamma_p(G_j)$$

Assume that q_{2ij} is finite for all i,j . Under these conditions, the system (1) is L_p-stable if the leading principal minors of $I_m - Q_2$ are all positive.

<u>Proof</u> Obvious from Theorem (12), if one observes that $\gamma_p(H_{ij} G_j) \leq \gamma_p(H_{ij}) \, \gamma_p(G_j)$. □

As before, one can take $q_{2ij} = a_{ij} \, b_j$, where a_{ij} is any <u>upper bound</u> for $\gamma_p(H_{ij})$ and b_j is any <u>upper bound</u> for $\gamma_p(G_j)$.

Now we consider the stability of the system (1) when some of the operators G_i and H_{ij} are perturbed. These results also show why Theorem (12) and Theorem (43) are called "weak interconnection" type.

45 Theorem Consider the system (1), and suppose that all the hypotheses of Theorem (12) are satisfied. Under these conditions, the system (1) remains L_p-stable whenever H_{ij} is replaced by $\alpha_{ij} H_{ij}$ for some $\alpha_{ij} \in [-1,1]$, $\forall i,j$.

Proof By assumption, we have that all the leading principal minors of $I_m - Q_1$ are positive. Now suppose we replace each H_{ij} by $\alpha_{ij} H_{ij}$, where $\alpha_{ij} \in [-1,1]$. Let $\bar{Q}_1 \in R^{m\times m}$ be defined by

46 $$\bar{q}_{1ij} = \gamma_p(\alpha_{ij} H_{ij} G_j)$$

Then, since

47 $$\gamma_p(\alpha_{ij} H_{ij} G_j) = |\alpha_{ij}| \gamma_p(H_{ij} G_j)$$

we have

48 $$0 \leq \bar{q}_{1ij} \leq q_{1ij}, \forall i,j$$

Since q_{1ij} is an upper bound for $\bar{q}_{1ij}$, the fact that all the leading principal minors of $I_m - Q_1$ are positive insures the L_p-stability of the perturbed system as well. □

49 Theorem Consider the system (1), and suppose that all the hypotheses of Theorem (43) are satisfied. Then the system (1) remains L_p-stable whenever each H_{ij} is replaced by $\alpha_{ij} H_{ij}$, each G_i is replaced by $\beta_i G_i$, and $\alpha_{ij} \in [-1,1]$, $\beta_i \in [-1,1]$.

Proof Entirely similar to that of Theorem (45). □

The properties brought out in Theorems (45) and (49) are usually referred to as connective stability [Sil. 1]. Theorem (45) states that if the system (1) is L_p-stable by virtue of satisfying the conditions of Theorem (12), then it remains stable if each interconnection operator H_{ij} is replaced by a "weaker" interconnection operator $\alpha_{ij} H_{ij}$ where $\alpha_{ij} \in [-1,1]$. In particular, the system remains stable even if we remove some previously present interconnections (i.e. let $\alpha_{ij} = 0$ for some i,j). Similar remarks apply to Theorem (49). Also, note that the condition that all the (leading) principal minors of $I_m - Q_1$ are positive automatically requires that $q_{1ii} < 1$ ∀i, i.e., that all the "selfinteractions" must be less than one (similarly Corollary (43) requires that $q_{2ii} < i$ ∀i). Finally, note that, given <u>any</u> $G_1,\ldots,G_m$ such that $\gamma_p(G_i) < \infty$ ∀i, one can <u>always</u> satisfy the criteria of corollary (43) (and hence of Theorem (12)) by making $\gamma_p(H_{ij})$ sufficiently small ∀i,j . This means that, if the system (1) is L_p-stable with $H_{ij} = 0$ ∀i,j, then it remains L_p-stable whenever $\gamma_p(H_{ij})$ is sufficiently small ∀i,j . For this reason, Theorem (12) and corollory (43) can be said to contain "weak interconnection" or "weak coupling" type of stability criteria. Also, the criteria of Theorem (12) and Corollary (43) are "sign-insensitive", in the sense that the test matrices Q_1 and Q_2 are unaffected even if we replace H_{ij} by $-H_{ij}$ for some i,j . It is intuitively clear that any stability criteria that can insure such a sturdy type of stability must necessarily be extremely conservative.

In the case of L_2-stability, it is possible to obtain a "sign-sensitive" criterion that contains Theorem (12) as a special case.

50 <u>Theorem</u> Consider the system (1), and suppose that p = 2 . Suppose that $\gamma_2(G_i) < \infty$ ∀i, and that there exist finite (not necessarily positive) constants α_{ij}, β_{ij} such that

51 $$\langle e_i, H_{ij} G_j e_j \rangle_T \geq -\alpha_{ij} \, ||e_i||_{T2} ||e_j||_{T2} - \beta_{ij} \, ||e_i||_{T2}, \quad \forall T \geq 0 ,$$

$$\forall e_i \in L_{pe}^{n_i}, \; \forall e_j \in L_{pe}^{n_j} .$$

Under these conditions, the system (1) is L_2-stable if there exist positive constants $\lambda_1,\ldots,\lambda_m$ such that the matrix $S \in R^{m\times m}$ defined by

52 $$s_{ij} = \lambda_i \, \delta_{ij} - (\lambda_i \, \alpha_{ij} + \lambda_j \, \alpha_{ji})/2$$

is positive definite.

The proof depends on the following

53 Lemma Suppose $A \in R^{m\times m}$ is positive definite, that $b \in R^m$, and that $d \in R$. Then there exist constants γ and c, depending only on A and d, such that whenever $x \in R^m$ satisfies

54 $$x' A x \leq b' x + d$$

x also satisfies

55 $$||x|| \leq \gamma ||b|| + c$$

where $||.||$ in (55) denotes the Euclidean norm on R^m.

Proof of Lemma (53) Let λ be the smallest eigenvalue of A. Then, whenever $x \in R^m$ satisfies (54), x also satisfies

56 $$\lambda ||x||^2 \leq b' x + d \leq ||b|| \; ||x|| + d$$

Completing the square in (56) readily yields (55). □

Proof of Theorem (50) From the system equations (1), we get

57 $$||e_i||_T^2 = \langle e_i, e_i\rangle_T = \langle e_i, u_i\rangle_T - \sum_{j=1}^{m} \langle e_i, H_{ij} G_j e_j\rangle_T$$

$$\leq ||e_i||_{T2} \, ||u_i||_{T2} + \sum_{j=1}^{m} \alpha_{ij}$$

$$||e_i||_{T2} ||e_j||_{T2} + \beta_{ij} \, ||e_i||_{T2}$$

where in the last step we use Schwarz's inequality and (52). Now, choose positive constants $\lambda_1, \ldots, \lambda_m$ such that matrix S is positive definite. Then from (57), we get

58 $$\sum_{i=1}^{m} \lambda_i \, ||e_i||_{T2}^2 \leq \sum_{i=1}^{m} \lambda_i \, ||e_i||_{T2}(||u_i||_{T2}+\beta_i) + \sum_{i=1}^{m} \sum_{j=1}^{m} \lambda_i \alpha_{ij} ||e_i||_{T2} ||e_j||_{T2}$$

59 $$\sum_{i=1}^{m} \sum_{j=1}^{m} (\lambda_i \delta_{ij} - \lambda_i \alpha_{ij}) \, ||e_i||_{T2} \, ||e_j||_{T2} \leq \sum_{i=1}^{m} ||e_i||_{T2}(\lambda_i ||u_i||_{T2} + \lambda_i \beta_i)$$

where

60 $$\beta_i = \sum_{j=1}^{m} \beta_{ij}$$

Suppose now that $u_i \in L_2^{n_i} \; \forall i$, and define $v_T \in R^m$, $b \in R^m$, and $M \in R^{m \times m}$ by

61 $$v_T = [||e_1||_{T2} \cdots ||e_m||_{T2}]'$$

62 $$b = [||u_1||_2 + \beta_1 \cdots ||u_m||_2 + \beta_m]'$$

63 $$M = \begin{bmatrix} \lambda_1(1-\alpha_{11}) & -\lambda_1 \alpha_{12} & \cdots & -\lambda_1 \alpha_{1m} \\ \vdots & & & \\ -\lambda_m \alpha_{m1} & -\lambda_m \alpha_{m2} & \cdots & \lambda_m(1-\alpha_{mm}) \end{bmatrix}$$

Then (59) is equivalent to

64 $$v_T' \, M \, v_T \leq b' \, v_T$$

Now note that S is the symmetric part of M, so that $v_T' \, M \, v_T = v_T \, S \, v_T$. Hence (64) is equivalent to

65 $$v_T' \, S \, v \leq b' \, v_T$$

Since S is positive definite, by Lemma (53) there exists a constant γ such that

66 $$\|v_T\| \le \gamma \, \|b\|$$

It is clear from (66) that $e_i \in L_2^{n_i}$ $\forall i$. Since $\gamma_2(G_i) < \infty$ $\forall i$, it follows that $y_i \in L_2^{n_i}$ $\forall i$. Finally, it is easy to see, from (66) and the fact that $\gamma_2(G_i) < \infty$ $\forall i$, that (24) holds in this case. Hence the system (1) is L_2-stable. □

Note that the criterion of Theorem (50) is "sign-sensitive", in that the test matrix S is in general altered if we replace H_{ij} by $-H_{ij}$. Also, "connective stability" (see (49) <u>et seq</u>.) does not necessarily hold if the criterion of Theorem (50) is satisfied. This leads one to believe that Theorem (50) gives less conservative criteria than Theorem (12).

We now show that Theorem (50) contains Theorem (12) as a special case. As a prelude, observe that if $\gamma_2(H_{ij}G_j) < \infty$, then there exists a finite constant β_{ij} such that

67 $$\|H_{ij}G_je_j\|_{T2} \le \gamma_2(H_{ij}G_j)\,\|e_j\|_{T2} + \beta_{ij}\,,\ \forall T \ge 0\,,\quad \forall e_j \in L_{2e}^{n_j}$$

So by Schwarz's inequality, we have

68 $$\begin{aligned}\langle e_i, H_{ij}G_je_j\rangle_T &\ge -\|e_i\|_{T2}\,\|H_{ij}G_je_j\|_{T2} \\ &\ge -\gamma_2(H_{ij}G_j)\,\|e_i\|_{T2}\,\|e_j\|_{T2} \\ &\quad - \beta_{ij}\,\|e_i\|_{T2}\,, \\ &\forall T \ge 0\,,\ \forall e_i \in L_{2e}^{n_i}\,,\ \forall e_j \in L_{2e}^{n_j}\end{aligned}$$

Hence (51) is satisfied with $\alpha_{ij} = \gamma_2(H_{ij}G_j)$.

69 <u>Lemma</u> Consider the system (1), and suppose $\gamma_2(G_i) < \infty$, $\gamma_2(H_{ij}G_j) < \infty$, $\forall i,j$. Under these conditions, if the leading

principal minors of the matrix $I_m - Q_1$ are all positive (recall (13)), then there exist positive constants $\lambda_1, \ldots, \lambda_m$ such that the matrix S of (52) is positive definite, with $\alpha_{ij} = \gamma_2(H_{ij}G_j)$.

Proof If we define $\underset{\sim}{\Lambda} = \text{Diag}\{\lambda_1, \ldots, \lambda_m\}$, then

70 $$S = [\Lambda\,(I_m - Q_1) + (I_m - Q_1)'\,\Lambda]/2$$

The lemma now follows readily from Fact (11) . □

We conclude this subsection by presenting a result similar to Corollary (43), but with a slightly different test matrix.

71 Theorem Consider the system (1), and suppose $\gamma_p(G_i) < \infty$, $\gamma_p(H_{ij}) < \infty$, $\forall i,j$. Define the matrix $Q_3 \in R^{m\times m}$ by

72 $$q_{3ij} = \gamma_p(G_i)\ \gamma_p(H_{ij})$$

Under these conditions, the system (1) is L_p-stable if the leading principal minors of $I_m - Q_3$ are all positive.

Proof Let the constants b_i, d_{ij} be chosen such that

73, 74 $$\left.\begin{aligned} ||G_jx||_T &\le \gamma_p(G_j)\ ||x||_T + b_j \\ ||H_{ij}x||_T &\le \gamma_p(H_{ij})\ ||x||_T + d_{ij} \end{aligned}\right\}\ \forall T \ge 0,\ \forall x \in L_{pe}^{n_j}$$

From (1), we get

75 $$\begin{aligned} ||y_i||_{Tp} &\le \gamma_p(G_i)\ ||e_i||_{Tp} + b_i \\ &\le \gamma_p(G_i)\ [||u_i||_{Tp} \\ &\quad + \sum_{j=1}^{m} \gamma_p(H_{ij})\ ||y_j||_{Tp} + d_{ij}] + b_i \end{aligned}$$

76 $$\sum_{j=1}^{m} (\delta_{ij} - q_{3ij})\ ||y_j||_{Tp} \le \gamma_p(G_i)\ ||u_i||_{Tp} + c_i$$

where

77 $$c_i = b_i + \gamma_p(G_i) \sum_{j=1}^{m} d_{ij}$$

The inequality (76) is similar to (19), and the proof of L_p-stability now follows along the same lines as that of Theorem (12). The details are left to the reader. □

Theorem (71) is not as good a result as Theorem (12), because the latter involves terms such as $\gamma_p(H_{ij}\ G_j)$, whereas the former involves terms such as $\gamma_p(G_i)\ \gamma_p(H_{ij})$, which is in general larger than $\gamma_p(G_i H_{ij})$. In this respect, Theorem (71) is comparable to Corollary (43).

We leave the proof of the following "connective stability" result to the reader, because it is entirely similar to that of Theorem (45).

78 Theorem Consider the system (1), and suppose that all the hypotheses of Theorem (71) are satisfied. Then the system (1) remains L_p-stable whenever each H_{ij} is replaced by $\alpha_{ij} H_{ij}$, each G_i is replaced by $\beta_i\ G_i$, and $\alpha_{ij} \in [-1,1]$, $\beta_i \in [-1,1]$.

6.2.3 CRITERIA BASED ON LOOP TRANSFORMATIONS

We now present some "loop-transformation" results that generalize Theorem (6.1.7). To do this, we need a

79 Lemma Consider a system described in aggregate form by

80 $$e = u - Hy \ , \ y = Ge$$

where $e,u,y \in L^n_{pe}$, and H,G map L^n_{pe} into itself. Suppose C and D are linear operators on L^n_{pe} such that $CD = DC = 0$, and define

81 $$f = e + Cy \ , \quad z = y - De$$

Suppose H is linear, and that $(I + HD)^{-1}$, $(I + CG)^{-1}$ exist as operators on L^n_{pe} . Under these conditions, the system equations (79) can be equivalently expressed in terms of the new variables z and f as

82a $$z = (G-D)(I+CG)^{-1}f$$

82b $$f = (I+HD)^{-1} u - (I+HD)^{-1}(H-C)z$$

<u>Proof</u> First of all, from (81) and (80) we get

83 $$z = Ge - De = (G-D)e$$

Next, we get from (81) that

84 $$e = f - Cy = f - CGe$$

85 $$e = (I+CG)^{-1} f$$

Combining (83) and (85) gives

86 $$z = (G-D)(I+CG)^{-1} f$$

which is (82a). Next, we have

87 $$\begin{aligned} f &= e + Cy = u - Hy + Cy = u - (H-C)y \\ &= u - (H - C)(z + De) \\ &= u - (H - C)z - HDe \qquad \text{(because } CD = 0) \\ &= u - (H-C)z - HDf + HDCy \\ &= u - (H - C)z - HDf \qquad \text{(because } DC = 0) \end{aligned}$$

88 $$f = (I+HD)^{-1} u - (I+HD)^{-1} (H-C)z$$

which is (82b). □

89 <u>Theorem</u> Consider a system described by

90a $$e_i = u_i - \sum_{j=1}^{m} H_{ij} y_j$$

90b $$y_i = G_i e_i$$

where $e_i, u_i, y_i \in L_{pe}^{n_i}$ for some integer n_i and some fixed $p \in [1,\infty]$, $G_i : L_{pe}^{n_i} \to L_{pe}^{n_i}$, and $H_{ij} : L_{pe}^{n_j} \to L_{pe}^{n_i}$. Define G and H as in (31) and (32). Select C and D of the form

91a $$C = \text{Diag } \{C_1,\ldots,C_m\}$$

91b $$D = \text{Diag } \{D_1,\ldots,D_m\}$$

such that all the hypothesis of Lemma (79) are satisfied. Define

92 $$F = (G-D)(I+CG)^{-1} = \text{Diag } \{(G_i-D_i)(I+C_iG_i)^{-1} \text{ , } i=1,\ldots,m\}$$

93 $$R = (I+HD)^{-1} (H-C)$$

and partition R in a manner compatible with that of H. Suppose $\gamma_p(F) < \infty$, $\gamma_p(R) < \infty$, and $\gamma_p[(I+HD)^{-1}] < \infty$, and define $Q_4 \in R^{m\times m}$ by

94 $$q_{4ij} = \gamma_p[R_{ij} (G_j-D_j) (I+C_jG_j)^{-1}]$$

Under these conditions, the system (90) is L_p-stable if the leading principal minors of I_m-Q_4 are all positive.

The proof is omitted, as the theorem follows readily by applying Theorem (12) to the modified system description (82).

Note that, with C and D of the form (91), the assumption that $CD = DC = 0$ means that $C_iD_i = 0 \; \forall i$. In particular, this assumption can be satisfied by setting either $C_i = 0$ or $D_i = 0$ for each i. In this case, the transformation (81) means that we either place a "feed-forward" of D_i or a "feedback" of C_i around the subsystem G_i, but never both.

To show that Theorem (6.1.7) is a special case of Theorem (89), consider once again the system (6.1.2), and let a linear operator $K : L_{pe}^{\nu} \to L_{pe}^{\nu}$ with $\gamma_p(K) < \infty$ be given. Select

95 $$C = \begin{bmatrix} K & 0 \\ 0 & 0 \end{bmatrix}, \quad D = \begin{bmatrix} 0 & 0 \\ 0 & K \end{bmatrix}$$

since

96 $$G = \begin{bmatrix} G_1 & 0 \\ 0 & G_2 \end{bmatrix}, \quad H = \begin{bmatrix} 0 & I \\ -I & 0 \end{bmatrix}$$

it is routine to verify that

97 $$(G-D)(I+CG)^{-1} = \begin{bmatrix} G_1(I+KG_1)^{-1} & 0 \\ 0 & G_2-K \end{bmatrix}$$

98 $$(I+HD)^{-1}(H-C) = H$$

99 $$(I+HD)^{-1} = \begin{bmatrix} I & -K \\ 0 & I \end{bmatrix}$$

Thus, all the conditions needed to apply Theorem (89) are satisfied, and the test matrix Q_3 becomes

100 $$Q_4 = \begin{bmatrix} 0 & \gamma_p(G_2-K) \\ \gamma_p[G_1(I+KG_1)^{-1}] & 0 \end{bmatrix}$$

From Theorem (89), the system (6.1.2) is L_p-stable if the leading principal minors of I_2-Q_4 are both positive, which is the case if

101 $$\gamma_p(G_2-K)\ \gamma_p[G_1(I+KG_1)^{-1} < 1$$

This is what Theorem (6.1.7) states.

6.3 CRITERIA BASED ON AN ESSENTIAL SET DECOMPOSITION

In this section, we present some criteria for L_p-stability based graph-theoretic techniques. The key concept here is that of an essential set of a graph. The criteria presented here generalize , in a natural way, Theorem (6.1.4) as well as the loop-transformation theorems (6.1.7) and (6.1.9).

Throughout this section, we study systems described by

1 $$e_i = u_i - \sum_{j=1}^{m} S_{ij}\, e_j \ , \quad i = 1,\ldots,m$$

where e_i, $u_i \in L_{pe}^{n_i}$ for some fixed $p \in [1,\infty]$ and some positive integer n_i , and $S_{ij} : L_{pe}^{n_j} \rightarrow L_{pe}^{n_i}$. For a discussion of the relationship between the system descriptions (1) and (6.2.1), see Section 2.2. Associated with the system (1), we construct the system digraph $\mathcal{D}$ as follows: Label m vertices as $v_1,\ldots,v_m$, and construct an edge (v_i,v_j) if $S_{ji} \neq 0$. If $\mathcal{D}$ is not strongly connected, we can find the strongly connected components of $\mathcal{D}$, as detailed in Section 4.2 . Hence we can assume without loss of generality that $\mathcal{D}$ is strongly connected.

We now define an essential set of D .

2 Definition Let $V = \{v_1,\ldots,v_m\}$ denote the vertex set and E the set of edges of $\mathcal{D}$. Given any subset V_2 of V , the section graph $\mathcal{D}_{V_2}$ is defined as the digraph whose edge set is $E \cap [(V-V_2)\times(V-V_2)]$.

In other words, the section graph $\mathcal{D}_{V_2}$ is obtained from $\mathcal{D}$ by (i) removing from V all the vertices in V_2 , and (ii) removing from E all the edges that either leave or enter a vertex in V_2 .

3 <u>Definition</u> A set of vertices $V_2 \subseteq V$ is called an essential set of $\mathcal{D}$ if the section graph $\mathcal{D}_{V_2}$ is acyclic (however, $\mathcal{D}_{V_2}$ is permitted to contain self-loops).

Let $A \in R^{m \times m}$ denote the adjacency matrix of $\mathcal{D}$, defined by $a_{ij} = 1$ if $(v_i, v_j) \in E$ and $a_{ij} = 0$ if $(v_i, v_j) \in E$. Similarly, let $\bar{A}$ denote the adjacency matrix of $\mathcal{D}_{V_2}$. Now, if $\mathcal{D}_{V_2}$ is acyclic (i.e. if V_2 is an essential set), then $\bar{A}$ can be made into a lower triangular matrix by a symmetric permutation of rows and columns (which corresponds to renumbering the vertices in $V-V_2$).

To see how this concept can be used in analyzing the system (1), let $V_2 \subseteq V$ be an essential set of the system diagraph $\mathcal{D}$, and let $V_1 = V-V_2$. Number the vertices in V_1 in such a way that the adjacency matrix of $\mathcal{D}_{V_2}$ is lower-triangular, and number every vertex in V_2 higher than every vertex in V_1 . With this renumbering, the adjacency matrix A of $\mathcal{D}$ is of the form

4 $$A = \begin{array}{c} \begin{matrix} k & \quad m-k \end{matrix} \\ \begin{bmatrix} A_{11} & A_{12} \\ A_{21} & A_{22} \end{bmatrix} \end{array} \begin{matrix} k \\ m-k \end{matrix}$$

where k is the number of vertices in V_1 , and A_{11} is lower-triangular. If we renumber the unknowns $e_1, \ldots, e_m$ in the same way, the system equations (1) become

5a $$e_i = u_i - \sum_{j-1}^{i} S_{ij} e_j - \sum_{j=k+1}^{m} S_{ij} e_j \ , \quad i = 1, \ldots, k$$

5b $$e_i = u_i - \sum_{j=1}^{m} S_{ij} e_j \ , \quad i = k+1, \ldots, m$$

Let us aggregate the equations (5) by defining

6a $$R_{11} = (S_{ij}) \ , \quad i = 1,\ldots,k \ ; \ j = 1,\ldots,k$$

6b $$R_{12} = (S_{ij}) \ , \quad i = 1,\ldots,k \ ; \ j = k+1,\ldots,m$$

6c $$R_{21} = (S_{ij}) \ , \quad i = k+1,\ldots,m \ ; \ j = 1,\ldots,k$$

6d $$R_{22} = (S_{ij}) \ , \quad i = k+1,\ldots,m \ ; \ j = k+1,\ldots,m$$

7a $$d_1 = (e_i), \quad i = 1,\ldots,k$$

7b $$d_2 = (e_i), \quad i = k+1,\ldots,m$$

8a $$v_1 = (v_i), \quad i = 1,\ldots,k$$

8b $$v_2 = (v_i), \quad i = k+1,\ldots,m$$

With these definitions, the system equations (5) become

9a $$d_1 = v_1 - R_{11} d_1 - R_{12} d_2$$

9b $$d_2 = v_2 - R_{21} d_1 - R_{22} d_2$$

where the operator $I+R_{11}$ is easy to invert because of the "triangular" nature of R_{11}. This is the point of the renumbering.

From the above discussion, it is clear that we would like to make the number k (i.e. the set V_1) as large as possible, by choosing as V_2 an essential set containing a minimum number of vertices, i.e. a so-called minimum essential set. In general, finding a minimum essential set of a given digraph is an NP-complete problem, and the computational complexity of known algorithms to solve this problem increases faster than any polynomial in m (the number of vertices of the digraph). In practice, however, there are efficient algorithms for finding "near-minimal" essential sets.

Once the system equations have been rearranged in the form (9), we can derive small-gain type stability criteria that generalize Theorems (6.1.1) and (6.1.4).

10 **Theorem** Consider the system (9), and suppose $(I+R_{11})^{-1}$ is well-defined. Suppose $\gamma_p(R_{ij}) < \infty\ \forall i,j$, and $\gamma_p((I+R_{11})^{-1}) < \infty$. Under these conditions, the system (9) is L_p-stable if

11 $$\gamma_p(R_{21}(I+R_{11})^{-1}) \cdot \gamma_p(R_{12}) + \gamma_p(R_{22}) < 1$$

Proof We can rewrite (9) in the form

12 $$d_1 = (I+R_{11})^{-1}(v_1 - R_{12}d_2)$$

13 $$d_2 = v_2 - R_{21}(I+R_{11})^{-1}(v_1 - R_{12}d_2) - R_{22}d_2$$

Suppose now that (11) holds, and that $v_1, v_2 \in L_p$. Then, taking norms in (13), we get

14 $$||d_2||_{Tp} \leq ||v_2||_p + \gamma_p(R_{21}(I+R_{11})^{-1})\ (||v_1||_p + \gamma_p(R_{12})$$
$$||d_2||_{Tp} + \gamma_p(R_{22})\ ||d_2||_{Tp} + \beta$$

where β is a finite constant. Rearranging (13), we get

15 $$||d_2||_{Tp} \leq [1-\gamma_p(R_{21}(I+R_{11})^{-1}) \cdot \gamma_p(R_{12}) - \gamma_p(R_{22})]^{-1}$$
$$[||v_2||_p + \gamma_p(R_{21}(I-R_{11})^{-1}\ ||v_1||_p + \beta]$$

It is now routine to verify that the system (9) is L_p-stable.□

Remarks In order to apply Theorem (10), we can take d_1 and d_2 to correspond to any arbitrary partitioning of the original set of unknowns $e_1,\ldots,e_m$. However, by choosing d_2 to correspond to an essential set of vertices, we make it very easy to verify whether or not $(I+R_{11})^{-1}$ exists. For instance, if $S_{ii} = 0$ in (1) whenever $v_i \in V_1$, then this assumption is automatically satisfied.

16 **Theorem** Consider the system (9), and suppose $(I+R_{11})^{-1}$, $(I+R_{22})^{-1}$ are well-defined. Suppose $\gamma_p(R_{ij}) < \infty\ \forall i,j$, and $\gamma_p((I+R_{ii})^{-1}) < \infty\ \forall i$. Under these conditions, the system (9) is L_p-stable if

17 $$\gamma_p((I+R_{22})^{-1}) \cdot \gamma_p(R_{21}(I+R_{11})^{-1}) \cdot \gamma_p(R_{12}) < 1$$

Proof In this case, (13) is equivalent to

18 $$d_2 = (I+R_{22})^{-1} [v_2 - R_{21}(I+R_{11})^{-1} (v_1 - R_{12}d_2)]$$

The rest of the proof is similar to that of Theorem (10). □

Theorems (10) and (16) both reduce to Theorem (6.1.1) in the case where $R_{11} = R_{22} = 0$, $R_{12} = -G_2$, $R_{21} = G_1$, which corresponds to the single-loop system (6.1.2). The next two results generalize Theorem (6.1.4). Their proofs are omitted as they are quite straightforward.

19 Theorem Suppose the conditions of Theorem (10) hold. Suppose in addition that either $v_2 = 0$ or that $R_{21}(I+R_{11})^{-1}$ is linear. Under these conditions, the system (9) is L_p-stable if

20 $$\gamma_p[R_{21}(I+R_{11})^{-1} R_{12} - R_{22}] < 1$$

21 Theorem Suppose the conditions of Theorem (16) hold. Suppose in addition that either $v_2 = 0$ or that $R_{21}(I+R_{11})^{-1}$ is linear. Under these conditions, the system (9) is L_p-stable if

22 $$\gamma_p((I+R_{22})^{-1}) \cdot \gamma_p(R_{21}(I+R_{11})^{-1} R_{12}) < 1 .$$

We now present some generalizations of the loop-transformation theorems (6.1.7) and (6.1.9).

23 Theorem Consider the system (1), and suppose $(I+R_{11})^{-1}$ is a well-defined linear operator. Suppose $\gamma_p(R_{ij}) < \infty \ \forall i,j$, and that $\gamma_p((I+R_{11})^{-1}) < \infty$. Under these conditions, the system (1) is L_p-stable if we can find a linear operator K with $\gamma_p(K) < \infty$ such that $(I+KR_{21})^{-1}$ exists, and

24 $$\gamma_p(R_{21}(I+KR_{21})^{-1}) \cdot \gamma_p[(I+R_{11})^{-1} R_{12} + K(R_{22}+I)] + \gamma_p(R_{22}) < 1$$

Proof Since K is linear, we have from (9b) that

25 $$Kd_2 = Kv_2 - KR_{21}\, d_1 - KR_{22}\, d_2$$

Substituting from (25) into (12), and using the linearity of $(I+R_{11})^{-1}$ gives

26 $$d_1 = (I+R_{11})^{-1} v_1-(I+R_{11})^{-1} R_{12}\, d_2 - K\, d_2 + K\, d_2$$

$$= (I+R_{11})^{-1} v_1-[(I+R_{11})^{-1} R_{12}+K]\, d_2 + Kv_2 - KR_{21}\, d_1 - KR_{22}\, d_2$$

27 $$d_1 = (I+KR_{21})^{-1} [(I+R_{11})^{-1} v_1+Kv_2 - ((I+R_{11})^{-1}R_{12} + K(R_{22}+I))d_2]$$

$$= (I+KR_{21})^{-1} [W-((I+R_{11})^{-1}R_{12} + K(R_{22}+I))d_2]$$

where $W = (I+R_{11})^{-1} v_1 + Kv_2 \in L_p$ whenever $v_1, v_2 \in L_p$. Substituting (27) into (9b) gives

28 $$d_2 = v_2-R_{21}(I+KR_{21})^{-1} [W-((I+R_{11})^{-1}R_{12}+I))d_2] - R_{22}d_2$$

The conclusion of the proof is now straight-forward. Taking norms in (28), and assuming that v_2, $w \in L_p$, we get

29 $$||d_2||_{Tp} \leq ||v_2||_p + \gamma_p(R_{21}(I+KR_{21})^{-1}) [||W|| + \gamma_p((I+R_{11})^{-1}R_{12}+K(R_{22}+1)) ||d_2||_{Tp}] + \gamma_p(R_{22}) ||d_2||_{Tp}$$

So the system is L_p-stable if (24) holds.

NOTES AND REFERENCES

The small gain theorem for feedback systems is due to Sandberg [San. 2] and Zames [Zam. 3]. It was extended to large-scale systems by Cook [Coo. 1] and Porter and Michel [Por. 1]. The connection between these results and the small gain theorem was established by Araki [Ara. 3]. The results on nonnegative matrices are due to Fiedler and Ptak [Fie. 1], Araki [Ara. 1] and Porter and Michel [Por. 1]. Additional results can be found in the work of Moylan [Moy. 1], and Siljak [Sil. 1]. The concept of connective stability is similar to that found in [Sil. 1]. The stability criteria based on an essential set decomposition are due to Callier, Chan and Desoer [Cal. 1]. See [Mic. 1] and [Las. 1] for some results not presented here.

CHAPTER 7: DISSIPATIVITY-TYPE CRITERIA FOR L_2-STABILITY

In this Chapter, we state and prove several "dissipativity-type" criteria for the stability of a large-scale interconnected system. These criteria differ in several ways from the "small-gain" type of criteria contained in Chapter 6. Most important, the basic dissipativity-type criteria are "sign-sensitive", whereas the basic small-gain type criteria (e.g. Theorems (6.2.12) and (6.3.10) are "sign-insensitive". As a result, whereas small-gain type criteria can be thought of as "weak interconnection" conditions, dissipativity-type criteria are in general less conservative and can be interpreted as "beneficial interconnection" conditions. On the other hand, dissipativity-type criteria require all the inputs, outputs, and errors to belong to an extended inner product space, whereas the small-gain criteria require only a normed space setting. As a consequence, small-gain criteria can be directly applied to study L_p-stability for all values of p, whereas dissipativity-type criteria can be directly applied to study only L_2-stability. In order to use dissipativity methods to derive L_∞-stability criteria, one has to resort to the technique of exponential weighting (see Chapter 9). Finally, it is interesting to note that there are many more distinct generalizations of the single-loop passivity theorem than there are of the single-loop small-gain theorem.

Throughout this chapter, we assume that each of the interconnection operators can be represented by a constant matrix of appropriate dimension, and use the symbol H to denote the interconnection matrix, as in (2.2.6). As noted previously (see Section 2.2), this assumption does not result in any loss of generality, and can always be satisfied by increasing the number of subsystems if necessary. Actually, we can routinely extend all of the results of this Chapter to the case where H_{ij} is a linear operator, but not necessarily a matrix. However, in this case, the stability "criteria" require a linear operator to be strongly passive (see Definition (3.2.16)). Since such a "criterion" involves verifying a property of the overall system, it would go against the general philosophy adopted here, which is to check the stability of the complex overall system by examining various subsystems and interconnections. Similarly, it is possible to extend the results here to the case where the H_{ij}'s are memoryless nonlinear operators; but we do not do this, for the same reason. Instead, we prefer to absorb all nonlinearities and dynamics into the subsystems, and assume that H_{ij} is a constant matrix, for all i,j. As a result, our stability criteria involve certain test matrices being positive definite or positive semidefinite..

Since we exclusively study L_2-stability in this chapter, we use $||\cdot||_T$ and $||\cdot||$ in place of $||\cdot||_{T2}$ and $||\cdot||_2$ respectively, in the interests of brevity.

7.1 SINGLE-LOOP SYSTEMS

In this section, we state and prove the basic dissipativity for single-loop systems. Even though these theorems are subsumed by subsequent results for large-scale systems, they are nevertheless of interest, both for their historical value as well as for their appealing physical interpretation.

Throughout this section, we consider a single-loop feedback system described by

1a $$e_1 = u_1 - y_2$$

1b $$e_2 = u_2 + y_1$$

1c $$y_1 = G_1 e_1$$

1d $$y_2 = G_2 e_2$$

where $u_1, u_2, e_1, e_2, y_1, y_2$ all belong to the space L_{2e}^{ν} for some positive integer ν, and G_1, G_2 map L_{2e}^{ν} into itself.

Recall now the following result from Chapter 6.

2 <u>Lemma</u> Suppose $A \in R^{n\times n}$ is positive definite, $b \in R^n$, and that $d \in R$. Then there exist constants γ and c, depending only on A and d, such that whenever $x \in R^n$ satisfies

3 $$x' A x \leq b' x + d$$

x also satisfies

4 $$||x|| \leq \gamma ||b|| + c$$

where $||.||$ denotes the Euclidean norm on R^n.

<u>Proof</u> See Lemma (6.2.53).

We now present the most general result of this section.

5 <u>Theorem</u> Suppose there exist real constants ε_i, δ_i, α_i, $i = 1,2$, such that the operators G_1 and G_2 satisfy the conditions

6 $$\langle x, G_i x\rangle_T \geq \varepsilon_i ||x||_T^2 + \delta_i ||G_i x||_T^2 + \alpha_i \quad \forall T \geq 0 \, , \; \forall x \in L_{2e}^{\nu} \, , \; i=1,2$$

Then the system (1) is L_2-stable, provided

7a $$\varepsilon_1 + \delta_2 > 0$$

7b $$\varepsilon_2 + \delta_1 > 0$$

<u>Proof</u> From the system equations (1), we have

8 $$\begin{aligned} \langle y_1, u_1\rangle_T + \langle y_2, u_2\rangle_T &= \langle y_1, e_1 + y_2\rangle_T + \langle y_2, e_2 - y_1\rangle_T \\ &= \langle y_1, e_1\rangle_T + \langle y_2, e_2\rangle T \\ &\geq \varepsilon_1 ||e_1||_T^2 + \delta_1 ||y_1||_T^2 + \alpha_1 + \varepsilon_2 ||e_2||_T^2 \\ &\quad + \delta_2 ||y_2||_T^2 + \alpha_2 \end{aligned}$$

Substituting for e_1 from (1a) and e_2 from (1b) gives

9 $$\langle y_1,u_1\rangle_T + \langle y_2,u_2\rangle_T \geq \varepsilon_1||u_1-y_2||_T^2 + \delta_1||y_1||_T^2 + \alpha_1 + \varepsilon_2||y_2+y_1||_T^2 + \delta_2||y_2||_T^2 + \alpha_2$$

$$= \varepsilon_1[||u_1||_T^2 - 2\langle u_1,y_2\rangle_T + ||y_2||_T^2] + \delta_1||y_1||_T^2 + \alpha_1 + \varepsilon_1[||u_2||_T^2 + 2\langle u_2,y_1\rangle_T + ||y_1||_T^2] + \delta_2||y_2||_T^2 + \alpha_2$$

Suppose now that $u_1, u_2 \in L_2^\nu$. Applying Schwarz's inequality repeatedly in (9), collecting all the quadratic terms in y_1, y_2 on the left, and the remaining terms on the right, gives

10 $$(\varepsilon_2+\delta_1)||y_1||_T^2 + (\varepsilon_1+\delta_2)||y_2||_T^2 \leq ||y_1||_T(||u_1|| + 2|\varepsilon_2|\,||u_2||) + ||y_2||_T(||u_2|| + 2|\varepsilon_1|\,||u_1||) + |\varepsilon_1|\,||u_1||^2 + |\varepsilon_2|\,||u_2||^2 + \alpha_1 + \alpha_2$$

where we have replaced $||u_1||_T$ and $||u_2||_T$ by their upper bounds $||u_1||$ and $||u_2||$ respectively. Since the left side of (10) is a positive definite quadratic form in $||y_1||_T$, $||y_2||_T$ whenever (7) holds, it follows by Lemma (2) that $||y_1||_T$ and $||y_2||_T$ are bounded independently of T, i.e. $y_1, y_2 \in L_2^\nu$. From (1a) and (1b), this implies that $e_1, e_2 \in L_2^\nu$. Furthermore, it is clear from (10) that the coefficients of the linear terms in $||y_1||_T$, $||y_2||_T$ are themselves linear in $||u_1||$, $||u_2||$. Thus, with reference to Lemma (2), the vector $\underset{\sim}{b}$ has components that are linear in $||u_1||, ||u_2||$. Now, it is clear from these argument that, not only do y_1, y_2 belong to L_2^ν , but also, there exist finite constants γ and δ such that $||y_1||$, $||y_2|| \leq \gamma(||u_1|| + ||u_2||) + \delta$. Similar remarks apply to e_1, e_2 . Thus the system (1) is L_2-stable. □

Several interesting criteria can be obtained as special cases of Theorem (5). Chief among them is the following result, which is usually known as the passivity theorem (for feedback systems):

11 <u>Corollary</u> Suppose there exist real constants ε, ζ, δ such that

12 $$\langle x, G_1 x\rangle_T \geq \varepsilon \, ||x||_T^2 \; , \quad \forall T \geq 0 \; , \quad \forall x \in L_{2e}^{\nu}$$

13 $$||G_1 x||_T \leq \zeta \, ||x||_T \; , \quad \forall T \geq 0 \; , \quad \forall x \in L_{2e}^{\nu}$$

14 $$\langle x, G_2 x\rangle_T \geq \delta \, ||x||_T^2 \; , \quad \forall T \geq 0 \; , \quad \forall x \in L_{2e}^{\nu}$$

Then the system (1) is L_2-stable, provided

15 $$\varepsilon + \delta > 0$$

<u>Proof</u> Since (15) holds, pick $\theta > 0$ such that

16 $$\varepsilon - \theta + \delta > 0$$

Now, (12) and (13) together imply that

17 $$\langle x, G_1 x\rangle_T \geq \varepsilon \, ||x||_T^2 = (\varepsilon-\theta) \, ||x||_T^2 + \theta \, ||x||_T^2$$

$$\geq (\varepsilon-\theta) \, ||x||_T^2 + (\theta/\zeta^2) \, ||G_1 x||_T^2 \, , \quad \forall T \geq 0 \; , \quad \forall x \in L_{2e}^{\nu}$$

Now apply Theorem (5) with $\varepsilon_1 = \varepsilon-\theta$, $\delta_1 = \theta/\zeta^2$, $\varepsilon_2 = 0$, and $\delta_2 = \delta$. □

It is also possible to prove Corollary (11) with (13) relaxed to another condition of the form:

18 $$||G_1 x||_T \leq \gamma \, ||x||_T + b \; , \quad \forall T > 0 \; , \quad \forall x \in L_{2e}^{\nu}$$

However, in this case, a slightly more general version of Theorem (5) would be needed, which is not as elegant as Theorem (5). Since the additive constant b is, generally speaking, present only in the study of L_∞-stability, this slight loss of generality in no way diminishes the usefulness of Corollary (11).

If, in Corollary (11), we assume that $\varepsilon > 0$ and $\delta = 0$, then we get the following result, which is historically the first version of the passivity theorem.

19 <u>Corollary</u> Suppose there exist positive constants ε and ζ such that

20 $$\langle x, G_1 x \rangle_T \geq \varepsilon \, ||x||_T^2 \ , \quad \forall T \geq 0 \quad , \quad \forall x \in L_{2e}^{\nu}$$

21 $$||G_1 x||_T \leq \zeta \, ||x||_T \ , \quad \forall T \geq 0 \quad , \quad \forall x \in L_{2e}^{\nu}$$

Suppose G_2 satisfies

22 $$\langle x, G_2 x \rangle_T \geq 0 \ , \ \forall T \geq 0 \ , \ \forall x \in L_{2e}^{\nu}$$

Under these conditions, the system (1) is L_2-stable.

An operator G_2 satisfying (22) is called a <u>passive</u> operator, whereas an operator G_1 satisfying (28) with $\varepsilon > 0$ is called a <u>strictly passive</u> operator. Hence Corollary (19) states the following: Suppose G_1 is strictly passive and has finite gain with zero bias, and that G_2 is passive; then the system (1) is L_2-stable.

Actually, both Corollaries (11) and (19) hold with G_1 and G_2 interchanged throughout. Note that Theorem (5) imposes "symmetric" conditions on G_1 and G_2 .

There are two other corollaries of Theorem (5) that are worth mentioning. Both are easy to prove.

23 <u>Corollary</u> Suppose both G_1 and G_2 are strictly passive; i.e., suppose there exist positive constants ε_1 and ε_2 such that

24 $$\langle x, G_i x \rangle_T \geq \varepsilon_i ||x||_T^2 \quad , \quad \forall T > 0 \ , \ \forall x \in L_{2e}^{\nu} \ , \ i = 1,2$$

Then the system (1) is L_2-stable.

25 <u>Corollary</u> Suppose there exist positive constants δ_1 and δ_2 such that

26 $$\langle x, G_i x \rangle_T \geq \delta_i ||G_i x||_T^2 \ , \ \forall T \geq 0 \ , \ \forall x \in L_{2e}^{\nu} \ , \ i = 1,2$$

Then the system (1) is L_2-stable.

Note that there is no assumption of finite gain in Corollary (23), and that there is no assumption of strict passivity in Corollary (25). Also, the results of section 3.2 are very useful for determining the various constants ε and δ corresponding to a given operator G.

7.2 GENERAL DISSIPATIVITY-TYPE CRITERIA

In this section we state and prove some general dissipativity-type criteria for the L_2-stability of large-scale interconnected systems. Special cases of the general results given here, which are easier to apply, are proved in the next section.

Throughout this section, and indeed troughout the rest of this chapter, we shall be concerned with a system described by

1a $$e_i = u_i - \sum_{j=1}^{m} H_{ij}\, y_j$$

$$i = 1,\ldots,m$$

1b $$y_i = G_i\, e_i$$

where u_i, e_i, y_i all belong to $L_{2e}^{n_i}$ for some integer n_i, G_i maps $L_{2e}^{n_i}$ into itself, and H_{ij} is a constant matrix of dimension $n_i \times n_j$. Then matrix $H \in R^{n\times n}$ (where $n = \sum_{i=1}^{m} n_i$) defined by

2 $$H = \begin{bmatrix} H_{11} & \cdots & H_{1m} \\ \vdots & & \\ H_{m1} & \cdots & H_{mm} \end{bmatrix}$$

is called the interconnection matrix.

We begin with an obvious

3 **Lemma** Suppose G_i is (Q_i, R_i, S_i)-dissipative for $i = 1,\dots,m$. Then $G : L_{2e}^n \to L_{2e}^n$ defined by

4 $$G = \text{Diag}\,\{G_1,\dots,G_m\}$$

is (Q,R,S)-dissipative, where

5a $$Q = \text{Diag}\,\{Q_1,\dots,Q_m\}$$

5b $$R = \text{Diag}\,\{R_1,\dots,R_m\}$$

5c $$S = \text{Diag}\,\{S_1,\dots,S_m\}$$

Proof Apply Definition (3.2.1).

The next result relates the dissipativity of G with that of the interconnected system (1).

6 **Lemma** Consider the system (1), and define G,H by (4) and (2), respectively. Suppose G is (Q,R,S)-dissipative. Then the system (1) is dissipative with respect to the triplet $(\bar{Q},\bar{R},\bar{S})$, where

7a $$\bar{Q} = Q + H'RH - \frac{1}{2}(SH + H'S')$$

7b $$\bar{R} = R$$

7c $$\bar{S} = S - 2H'R$$

Proof The system equations (1) can be compactly written as

8 $$e = u - Hy\,, \quad y = Ge$$

Since G is (Q,R,S)-dissipative, we have

9 $$\langle y,Qy\rangle_T + \langle e,Re\rangle_T + \langle y,Se\rangle_T \geq 0\,, \quad \forall T \geq 0\,, \quad \forall e \in L_{2e}^n$$

Substituting for e from (8) into (9) gives

10 $\langle y,Qy\rangle_T + \langle u-Hy,Ru-RHy\rangle_T + \langle y,Su-SHy\rangle_T \geq 0 \ , \ \forall T \geq 0 \ , \ \forall u \in L_{2e}^n$

or, after expanding and collecting terms,

11 $\langle y,(Q+H'RH-SH)y\rangle_T + \langle u,Ru\rangle_T + \langle y,(S-2H'R)u\rangle_T \geq 0 \ , \ \forall T \geq 0 \ ,$

$\forall u \in L_{2e}^n$

Since u is the input and y is the output of the system (1), (11) shows that the system (1) is $(\bar{Q},\bar{R},\bar{S})$-dissipative. (Note that $\bar{Q}$ is the symmetric part of $Q+H'RH-SH$). □

The next theorem gives the basic dissipativity-type criterion for L_2-stability.

12 Theorem Consider the system (1), and suppose G is (Q,R,S) dissipative. Then the system (1) is L_2-stable if $\bar{Q}$ of (7a) is negative definite.

Proof By Lemma (3.2.10), if $\bar{Q}$ is negative definite, then the relation mapping u into y has finite gain with zero bias. □

Remarks As was pointed out in Chapter 3, the concepts of gain and dissipativity can be defined for relations, though in Chapter 3 they are defined for operators to keep the exposition simple.

The next result gives a necessary and sufficient condition for the relation between u and y to have finite gain with zero bias. The necessity part is of little value as a testing procedure, but is useful to show that the condition of Theorem (12) is a reasonable one.

13 Theorem Consider the system (1), and let $F \subset L_{2e}^n \times L_{2e}^n$ denote the relation between u and y as defined by (1). Under these conditions, $\zeta_2(F) < \infty$ if and only if there exist matrices Q,R,S such that G is (Q,R,S)-dissipative and $\bar{Q}$ of (7a) is negative definite.

<u>Proof</u> "if" immediate from Theorem (12).

"only if" Suppose $\zeta_2(F) < \infty$. Then

14 $$-\langle y,y\rangle_T + \langle u,\zeta_2^2(F)u\rangle_T \geq 0 , \quad \forall u, y \in F$$

However, since u and y satisfy (1), we can replace u by e+Hy in (14). This gives

15 $$-\langle y,y\rangle_T + \langle e+Hy,\zeta_2^2(F)\ (e+Hy)\rangle_T \geq 0 , \quad \forall e \in L_{2e}^n , y = Ge$$

Expanding and collecting terms in (15) gives

16 $$\langle y,(\zeta_2^2(F)H'H-I_n)y\rangle_T + \langle e,\zeta_2^2(F)e\rangle_T + 2\langle y,\zeta_2^2(F)H'e\rangle_T \geq 0 ,$$
$$\forall e \in L_{2e}^n , y = Ge$$

This shows that G is (Q,R,S)-dissipative with

17 $$Q = \zeta_2^2(F)\ H'H-I_n , \ R = \zeta_2^2(F)I_n , \ \delta = 2\ \zeta_2^2(F)\ H'$$

Also, it is obvious from (14) that $\bar{Q} = -I_n$, which is negative definite. □

In order to give some flexibility in the application of Theorem (12), we can introduce some auxiliary constants. This gives the most general stability criterion of the dissipativity type, and all subsequent criteria are special cases of it.

18 <u>Theorem</u> Consider the system (1), and suppose G_i is (Q_i,R_i,S_i)-dissipative for $i = 1,\dots,m$. Under these conditions, the system (1) is L_2-stable if there exist positive constants $\lambda_1,\dots,\lambda_m$ such that the test matrix

19 $$M = -\ [\Lambda Q+H'\Lambda RH] + (\Lambda SH+H'S'\Lambda)/2$$

is positive definite, where

20 $$\Lambda = \mathrm{Diag}\ \{\lambda_1\ I_{n_1},\dots\lambda_m\ I_{n_m}\}$$

Proof Since G_i is (Q_i,R_i,S_i)-dissipative, we have

21 $$\langle G_ix_i,Q_iG_ix_i\rangle_T + \langle x_i,R_ix_i\rangle_T + \langle G_ix_i,S_ix_i\rangle_T \geq 0 ,$$
$$\forall T \geq 0 , \forall x_i \in L_{2e}^{n_i}$$

Multiplying (21) by λ_i , and summing as i varies from 1 to m shows that G of (4) is dissipative with respect to the triplet $(\Lambda Q,\Lambda R,\Lambda S)$ where Q,R,S are defined by (5). Now, applying Theorem (12), we get that $\bar{Q}$ of (7a) is in this case $-M$. Hence the system (1) is L_2-stable if M is positive definite. □

We conclude this section with three useful corollaries of Theorem (18).

22 Corollary Suppose that, for some choice of Λ , the matrix ΛSH is positive, i.e., $v'\Lambda SH\ v \geq 0\ \forall u \in R^n$.* Under these conditions, the system (1) is L_2-stable if the test marrix

23 $$M_o = -[\Lambda Q + H'\Lambda RH]$$

is positive definite.

Proof If ΛSH is positive, then $M \geq M_o$, so that if M_o is positive definite, so is M . □

24 Corollary Suppose H is nonsingular, and let $P = H^{-1}$. Then the system (1) is L_2-stable if there exist positive constants $\lambda_1,\ldots,\lambda_m$ such that the test matrix

25 $$N = -[P'\Lambda QP + \Lambda R] + (P'\Lambda S + S'\Lambda P)/2$$

is positive definite.

Proof Note that $N = P'MP$, so that $N > 0$ if and only if $M > 0$. □

---

*we reserve the term "positive semidefinite" for symmetric matrices.

26 **Corollary** Suppose that ΛSH is positive for some Λ, and that H is nonsingular. Then the system (1) is L_2-stable if the test matrix

27 $$N_o = -[P'\Lambda QP + \Lambda R]$$

is positive.

P **Proof** Note that $N_o = P'M_oP$, so that $N_o > 0$ if and only if $M > 0$. □

7.3 SPECIAL CASES: SMALL-GAIN AND PASSIVITY-TYPE CRITERIA

Theorem (7.12.18) is a powerful general result that can be applied to a wide variety of situations, e.g. when some of the operators G_i have finite gain, others are passive, still others are strictly passive, etc. By specializing Theorem (7.2.18), one can obtain several useful, and readily applicable, stability criteria. Some such results are presented in this section. As always, we study a system described by

1a $$e_i = u_i = \sum_{j=1}^{m} H_{ij}\, y_i$$

1b $$y_i = G_i\, e_i$$

$$i = 1,\ldots,m$$

where $u_i, e_i, y_i \in L_{2e}^{n_i}$ for some integer n_i, $G_i : L_{2e}^{n_i} \to L_{2e}^{n_i}$, and H_{ij} is a constant matrix of dimension $n_i \times n_j$.

First, we present a "small gain" type criterion based on the dissipativity approach.

2 **Theorem** Consider the system (1), and suppose $\zeta_2(G_i) < \infty$ $\forall i$. Define the test matrix $W \in R^{n\times n}$ by

---

*By $A \geq B$ $(A > B)$, we mean that $A-B$ is positive semi-definite (positive definite)

3 $$W = \begin{bmatrix} W_{11} & \cdots & W_{1m} \\ \vdots & & \\ W_{m1} & \cdots & W_{mm} \end{bmatrix}$$

4 $$W_{ij} = \zeta_2(G_i)\ H_{ij}$$

Under these conditions, the system (1) is L_2-stable if one can find positive constants $\lambda_1,\ldots,\lambda_m$ such that the matrix

5 $$\bar{M} = \Lambda - W'\Lambda W$$

is positive definite, where

6 $$\Lambda = \text{Diag}\ \{\lambda_1\ I_{n_2}\ ,\ldots,\lambda_m\ I_{n_m}\}$$

<u>Proof</u> Since $\zeta_2(G_i) < \infty$, we can write

7 $$\langle -G_i x, G_i x\rangle_T + \langle x, \zeta_i^2\ x\rangle_T \geq 0, \quad \forall T \geq 0, \quad \forall x \in L_{2e}^{n_i}$$

where we use $\zeta_i \triangleq \zeta_2(G_i)$ in the interests of brevity. Thus G_i is dissipative with respect to $(-I_{n_i}, \zeta_i^2\ I_{n_i}, 0)$. Applying Theorem (7.2.18), we get that the test matrix M of (7.2.19) is

8 $$M = \Lambda - H'\Lambda RH$$

where

9 $$R = \text{Diag}\ \{\zeta_1^2\ I_{n_1},\ldots,\zeta_m^2\ I_{n_m}\}$$

Now note that, since R is diagonal, we have $\Lambda R = R^{1/2}\Lambda\ R^{1/2}$. Hence

10 $$M = \Lambda - H'R^{1/2}\ \Lambda R^{1/2}\ H = \Lambda - W'\Lambda W = \bar{M}$$

Thus the system (1) is L_2-stable if Λ can be found such that $\bar{M}$ is positive definite. □

Note that, by the discrete-time version of Liapunov's theorem [Fre. 1, p.166], one can find a $\Lambda > 0$ such that $\bar{M} > 0$ if and only if $\rho(W) < 1$. However, even if $\rho(W) < 1$, one may not be able to find a diagonal $\Lambda > 0$ such that $\bar{M} > 0$. If we try to give a more explicit criterion than Theorem (2), i.e., one that does not depend on being able to find some unknown constants (with no systematic procedure to find them), we get a result that is very similar to Theorem (6.2.71).

11 Theorem Consider the system (1), and suppose $\zeta_2(G_i) < \infty$ $\forall i$. Define the test matrix $\bar{N} \in R^{m \times m}$ by

12 $$\bar{n}_{ij} = \zeta_2(G_i) \; ||H_{ij}|| = \zeta_i \; ||H_{ij}||$$

where $||H_{ij}||$ is the ℓ_2-induced norm of the matrix H_{ij}. Under these conditions, the system (1) is L_2-stable if all leading principal minors of the matrix $I-\bar{N}$ are positive.

The proof depends on the following lemma, which is proved in [Ara. 2].

13 Lemma Suppose a matrix $A \in R^{m \times m}$ has all non-negative elements, and that the leading principal minors of I_m-A are all positive. Under these conditions, there exists a diagonal matrix $\bar{\Lambda} > 0$ such that $\bar{\Lambda} - A'\bar{\Lambda} A > 0$.

As brought out in Lemma (6.2.8), the fact that $a_{ij} \geq 0$ and the nonnegativity of the leading principal minors of I_m-A implies that $\rho(A) < 1$. Thus, by Liapunov's theorem, there exists a $\bar{\Lambda} > 0$ such that $\bar{\Lambda}-A'\bar{\Lambda}A > 0$. The special property of A, namely $a_{ij} \geq 0$ $\forall i,j$, allows us to select $\bar{\Lambda}$ to be diagonal as well.

Proof of Theorem (11) By Lemma (13), if the hypotheses of Theorem (11) are satisfied, then there exists a diagonal $\bar{\Lambda} > 0$ such that $\bar{\Lambda} - \bar{N}' \bar{\Lambda} \bar{N} > 0$. Let $\lambda_1, \ldots, \lambda_m$ be the diagonal elements of this $\bar{\Lambda}$, and define Λ by (6). We claim that, with this choice, $\bar{M}$ of (5) is positive definite.

To establish this claim, note that $\bar{\Lambda}-\bar{N}'\bar{\Lambda}N > 0$

implies that there exists an $\alpha > 0$ such that

14 $$\sum_{i=1}^{m} \alpha v_i^2 \leq \sum_{i=1}^{m} \lambda_i v_i^2 - \sum_{i=1}^{m} \lambda_i \left(\sum_{j=1}^{m} \bar{n}_{ij} v_j\right)^2 , \quad \forall v \in R^m$$

Now let $x \in R^n$, and partition x as $[x_1' \ldots x_m']$, where $x_i \in R^{n_i}$. Then

15 $$\begin{aligned} x' \bar{M} x &= \sum_{i=1}^{m} \lambda_i ||x_i||^2 - \sum_{i=1}^{m} \lambda_i || \sum_{j=1}^{m} \zeta_i H_{ij} x_j ||^2 \\ &\geq \sum_{i=1}^{m} \lambda_i ||x_i||^2 - \sum_{i=1}^{m} \lambda_i \left(\sum_{j=1}^{m} \zeta_i ||H_{ij}|| \, ||x_j||\right)^2 \\ &= \sum_{i=1}^{m} \lambda_i ||x_i||^2 - \sum_{i=1}^{m} \lambda_i \left(\sum \bar{n}_{ij} ||x_j||\right)^2 \\ &\geq \alpha \sum_{i=1}^{m} ||x_i||^2 = \alpha x'x , \text{ by (14)} \end{aligned}$$

This shows that $\bar{M} > 0$. Thus, by Theorem (2), the system (1) is L_2-stable. □

The rest of this section is devoted to deriving passivity-type stability criteria. In what follows, we assume that each of the operators G_i satisfies one or the other of the following conditions: (i) there exist constants ε_i and ζ_i such that

16 $$\langle x, G_i x \rangle_T \geq \varepsilon_i ||x||_T^2$$

$$\forall T \geq 0 , \forall x \in L_{2e}^{n_i}$$

17 $$||G_i x||_T \leq \zeta_i ||x||_T$$

or (ii) there exists a constant δ_i such that

18 $$\langle x, G_i x \rangle_T \geq \delta_i ||G_i x||_T^2 , \quad \forall T \geq 0 , \forall x \in L_{2e}^{n_i}$$

Throughout, we also assume that the interconnection matrix H is positive, i.e.,

19 $$x' Hx \geq 0 \quad \forall x \in R^n$$

If H is positive, then from (1) we get

20 $$\sum_{i=1}^{m} \langle y_i, e_i \rangle_T = \sum_{i=1}^{m} \langle y_i, u_i \rangle_T - \sum_{i=1}^{m} \sum_{j=1}^{m} \langle y_i, H_i, y_j \rangle_T$$

$$\leq \sum_{i=1}^{m} \langle y_i, u_i \rangle_T$$

Thus the assumption that H is positive implies that the total energy absorbed by the m subsystems is no larger than the total energy entering through the inputs. In this case, by a slight abuse of language one can say that the system (1) has <u>passive interconnections</u>. Further, if H is skew-symmetric, then (20) holds with an equality, and we say that the system (1) has lossless interconnections. Note that the standard single-loop feedback system (7.1.1) has lossless interconnections.

We first prove two lemmas concerning matrices.

21 <u>Lemma</u> Let $A \in R^{\ell \times \ell}$ be positive semi-definite, and let $A_{22} \in R^{\ell_2 \times \ell_2}$ be positive definite, where

22 $$A = \begin{bmatrix} A_{11} & A_{12} \\ A'_{12} & A_{22} \end{bmatrix} \begin{matrix} \ell_1 \\ \ell_2 \end{matrix} \qquad (\text{column blocks } \ell_1, \ell_2)$$

Let $\Delta \in R^{\ell_1 \times \ell_1}$ be positive definite. Then the matrix $B \in R^{\ell \times \ell}$ defined by

23 $$B = \begin{bmatrix} A_{11} + \Delta & A_{12} \\ A'_{12} & A_{22} \end{bmatrix} = A + \begin{bmatrix} \Delta & 0 \\ 0 & 0 \end{bmatrix}$$

is positive definite.

Proof Clearly B is at least positive semidefinite, since it is the sum of two positive semidefinite matrices. Hence, to show that B is positive definite, it is enough to show that $x' B x = 0 \Rightarrow x = 0$. Let $x \in R^{\ell}$ and suppose $x' B x = 0$. Then $x' A x = 0$ and $x'_1 \Delta x_1 = 0$, where $x_1 \in R^{\ell_1}$ denotes the first ℓ_1 components of x. Now, since Δ is positive definite, $x_1' \Delta x_1 0 \Rightarrow x_1 = 0$. Hence $x' A x = x_2' A_{22} x_2 = 0$, where $x_2 \in R^{\ell_2}$ denotes the last ℓ_2 components of x. Since A_{22} is positive definite, this implies that $x_2 = 0$. Hence $x = 0$ and the lemma is proved.

24 Lemma Let $P, Q \in R^{\ell \times \ell}$ be positive semidefinite, and suppose

25 $$\text{rank } P = \text{rank } [P|Q]$$

Then there exists an $\alpha > 0$ such that $P - \alpha Q$ is positive semidefinite.

Proof The condition (9) implies that the range of Q is contained in the range of P, or equivalently, that the null space of Q contains the null space of P. Now, we can express $R^{\ell \times \ell}$ as a direct sum of the range of P and the null space of P, which are orthogonal subspaces. With this new basis, the matrix representation of P becomes

26 $$P \rightarrow \begin{array}{c} \\ \\ \\ \end{array} \begin{array}{cc} k & \ell-k \\ \left[\begin{array}{c} P_1 \\ \\ 0 \end{array}\right. & \left.\begin{array}{c} 0 \\ \\ 0 \end{array}\right] \end{array} \begin{array}{l} \\ k \\ \\ \ell-k \end{array}$$

where k is the rank of P, and P_1 is positive definite. Now, since the null space of Q contains that of P, the matrix representation of Q with respect to the same basis is of the form

27 $$Q \to \begin{bmatrix} Q_1 & 0 \\ 0 & 0 \end{bmatrix}$$

Since P_1 is positive definite, it is clear that $P_1 - \alpha Q_1$ is also positive definite for sufficiently small α. Hence $P-\alpha Q$ is positive semidefinite for sufficiently small α.

With the aid of these two lemmas, we can now state the stability criteria.

28 Theorem Consider the system (1). Suppose that for some integer $k \leq m$, there exist constants ε_i and ζ_i, $i=1,\ldots,k$, such that

29 $$\langle x, G_i x\rangle_T \geq \varepsilon_i ||x||_T^2 , \quad \forall T \geq 0 , \quad \forall x \in L_{2e}^{n_i}$$

$$i=1,\ldots,k$$

30 $$||G_i x||_T \leq \zeta_i \, ||x||_T , \quad \forall T \geq 0 , \quad \forall x \in L_{2e}^{n_i}$$

Suppose there exist constants δ_i, $i=k+1,\ldots,m$, such that

31 $$\langle x, G_i x\rangle_T \geq \delta_i ||G_i x||_T^2 , \quad \forall T \geq 0 , \quad \forall x \in L_{2e}^{n_i} , \quad i=k+1,\ldots,m$$

Let $r = \sum_{j=1}^{m} n_j$, and partition the interconnection matrix H as follows:

32 $$H = \begin{array}{c} \\ \begin{bmatrix} H_{aa} & H_{ab} \\ H_{ba} & H_{bb} \end{bmatrix} \end{array} \begin{array}{l} \\ r \\ n-r \end{array} \quad \text{(column blocks: } r, \; n-r\text{)}$$

Define diagonal matrices $E_a \in R^{r\times r}$ and $D_b \in R^{(n-r)\times(n-r)}$ by

33 $$E_a = \begin{bmatrix} \varepsilon_1 I_{n_1} & & 0 \\ & \ddots & \\ 0 & & \varepsilon_k I_{n_k} \end{bmatrix}$$

34 $$D_b = \begin{bmatrix} \delta_{k+1} I_{n_{k+1}} & & 0 \\ & \ddots & \\ 0 & & \delta_m I_{n_m} \end{bmatrix}$$

Under these conditions, the system (1) is L_2-stable if

(i) the matrix $M_1 \in R^{n\times n}$ defined by

35 $$M_1 = \begin{bmatrix} H'_{aa} E_a H_{aa} & H'_{aa} E_a H_{ab} \\ H'_{ab} E_a H_{aa} & H'_{ab} E_a H_{ab} + D_b \end{bmatrix}$$

is positive semidefinite.

(ii) the matrix $H'_{ab} E_a H_{ab} + D_b$ is positive definite.

(iii) we have

36 $$\text{rank } M_1 = \text{rank } \left[M_1 \,\middle|\, \frac{H'_{aa}}{H'_{ab}} \right]$$

(iv) the matrix H is positive.

<u>Proof</u> Let $Q \in R^{n\times n}$ be defined by

37 $$Q = \begin{bmatrix} H'_{aa} \\ \hline H'_{ab} \end{bmatrix} [H_{aa} | H_{ab}]$$

Then condition (36) states that the range of M_1 contains that of $\begin{bmatrix} H'_{aa} \\ \hline H'_{ab} \end{bmatrix}$, or equivalently, that the null space of $[H_{aa} | H_{ab}]$ contains that of M_1. Now, it is clear from (37) that the null space of Q is the same as that of $[H_{aa} | H_{ab}]$. Hence (36) implies that the null space of Q contains that of M_1. Hence by Lemma (24), there exist an $\alpha > 0$ such that $M_1 - \alpha Q$ is positive semidefinite.

Now pick $\alpha > 0$ sufficiently small such that $M_1 - \alpha Q$ is positive semidefinite and such that the matrix $H'_{ab} (E_a - \alpha I) H_{ab} + D_b$ is positive definite. The latter is always possible in view of condition (ii). From (29) and (30), it readily follows that for $i = 1, \ldots, k$,

38 $$\langle x, G_i x\rangle_T \geq \varepsilon_i ||x||_T^2 = (\varepsilon_1 - \alpha) \; ||x||_T^2 + \alpha \; ||x||_T^2$$
$$\geq (\varepsilon_i - \alpha) \; ||x||_T^2 + (\alpha/\zeta_i^2) \; ||G_i x||_T^2 \,, \quad \forall T \geq 0 \,,$$
$$\forall x \in L_{2e}^{n_i}$$

Hence, for $i = 1, \ldots, k$, the operator G_i is dissipative with respect to

39 $$Q_i = -(\alpha/\zeta_i^2) \; I_{n_i} \,, \; R_i = -(\varepsilon_i - \alpha) \; I_{n_i} \,, \; S_i = I_{n_i}$$

Similarly, from (31), we see that for $i = k+1, \ldots, m$, the operator G_i is dissipative with respect to

40 $$Q_i = -\delta_i \; I_{n_i} \,, \; R_i = 0_{n_i} \,, \; S_i = I_{n_i}$$

Let us now apply corollary (7.1.22) with $\Lambda = I_n$. In this case, $\Lambda SH = H$ is positive, so that corollary (7.1.22) is applicable. Forming the test matrix M_0, we get

41 $$M_0 = -[Q + H' RH]$$

$$= \begin{bmatrix} \Delta & 0 \\ 0 & D_b \end{bmatrix} + H' \begin{bmatrix} E_a - \alpha I_r & 0 \\ 0 & 0 \end{bmatrix} H$$

$$= M_1 - \alpha Q + \begin{matrix} \Delta & 0 \\ 0 & 0 \end{matrix}$$

where

42 $$\Delta = \text{Diag}\,\{\alpha/\zeta_1^2\, I_{n_1}, \ldots, \alpha/\zeta_k^2\, I_{n_k}\}$$

We claim that this matrix is positive definite. First of all, the first term, namely $M_1 - \alpha Q$, is positive semidefinite. Next, its submatrix $H'_{ab}\,(E_a - \alpha I_r)\, H_{ab} + D_b$ is positive definite. Finally, Δ is positive definite. Hence by Lemma (21), M_o is positive definite, whence by Corollary (7.2.22), the system (1) is L_2-stable. □

Remarks 1) Let us refer to an operator G_i satisfying (29) as pseudo-strictly passive ("pseudo" because there is no assumption that ε_i is positive). Similarly, an operator G_i satisfying (31) is said to be pseudo-passive. Clearly Theorem (28) gives a sufficient condition for the L_2-stability of an interconnection of some pseudo-strictly passive operators with finite gain and some pseudo-passive operators.

2) One of the interesting aspects of Theorem (28) is that it allows a "trade-off" between the constants ε_i and δ_i. This trade-off is one of the noteworthy features of the single-loop passivity theorem (Theorem (7.1.2)). In this context, it is clear from Theorem (28) that one should try to choose the constants ε_i and δ_i (as appropriate) to be as positive as possible in order to have the best chance of satisfying the stability conditions.

3) Note that if M_1 is positive definite then (ii) and (iii) are automatically satisfied.

4) Note that the hypotheses of Theorem (28) require the submatrix $H'_{aa} E_a H_{aa}$ to be positive semidefinite.

5) The submatrices H_{ba} and H_{bb} do not figure in the stability conditions of Theorem (28), except in condition (iv) (namely, the requirement that the interconnection matrix H be positive). Thus, for all practical purposes, the stability criteria depend only on the submatrices H_{aa} (the "self-interaction" among the pseudo-strictly passive subsystems) and H_{ab} (the interaction from the pseudo-passive subsystems to the pseudo-strictly passive subsystems). If conditions (i) thru (iii) of Theorem (28) are satisfied, and if H_{aa} is positive, then one can always add "compensation" to adjust H_{ba} and H_{bb} so as to make H positive, and thereby stabilize the system.

43 Example Consider an interconnection of three subsystems represented by the operator G_1, G_2, G_3, where G_1 pseudo-strictly passive and has finite gain, and G_2, G_3 are pseudo-passive. Let the interconnection matrix be

44 $$H = \begin{bmatrix} 0 & 1 & 0 \\ -1 & 0 & -1 \\ 0 & 1 & 0 \end{bmatrix}$$

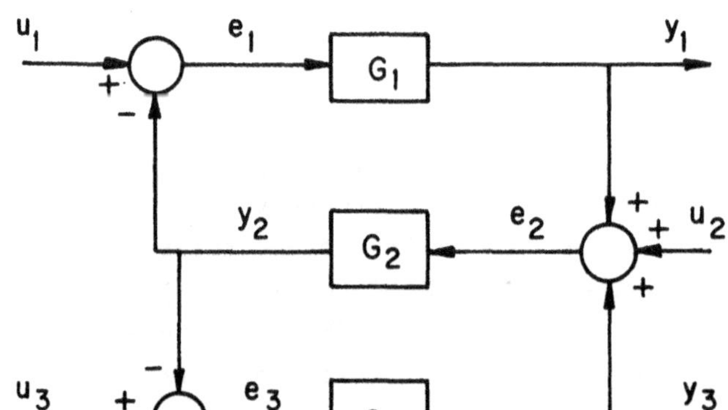

FIGURE 7.1

This system is shown in Figure 7.1. For this system, the matrix M_1 in (35) is

45 $$M_1 = \begin{bmatrix} 0 & 0 & 0 \\ 0 & \varepsilon_1+\delta_2 & 0 \\ 0 & 0 & \delta_3 \end{bmatrix}$$

Hence all the conditions of Theorem (28) are satisfied and the system at hand is L_2-stable if

46 $$\varepsilon_1+\delta_2 > 0 \ , \quad \delta_3 > 0$$

We now show that Theorem (28) contains Theorem (7.1.2) a special case. Consider the single-loop feedback system (7.1.1), and suppose G_1 , G_2 satisfy

47 $$<x,G_1x>_T \geq \varepsilon \ ||x||_T^2 \ , \quad \forall T > 0 \ , \quad \forall x \in L_{2e}^{\nu}$$

48 $$||G_1x||_T \leq \zeta \ ||x||_T \ , \quad \forall T \geq 0 \ , \quad \forall x \in L_{2e}^{\nu}$$

49 $$<x,G_2x>_T \geq \delta \ ||G_2x||_T^2 \ , \quad \forall T \geq 0 \ , \quad \forall x \in L_{2e}^{\nu}$$

Then the interconnection matrix is

50 $$\underset{\sim}{H} = \begin{bmatrix} 0_\nu & I_\nu \\ -I_\nu & 0_\nu \end{bmatrix}$$

Hence the test matrix M_1 of (35) becomes

51 $$M_1 = \begin{bmatrix} 0_\nu & 0_\nu \\ 0_\nu & (\varepsilon+\delta)I_\nu \end{bmatrix}$$

It is easy to verify that all conditions of Theorem (12) are satisfied if $\varepsilon+\delta > 0$.

We now present another stability criterion, which also contains the single-loop Theorem (7.1.2) as a special case, but is in general more restrictive than Theorem (28). However, it has the advantage that it has an instability counterpart (see Theorem (8.3.42)).

52 <u>Theorem</u> Consider the system (1), and suppose $G_1,\ldots,G_m$ satisfy (29)-(31), as appropriate. Suppose H is nonsingular and positive, and let $P = H^{-1}$. Partition P as

53
$$P = \begin{array}{c} \begin{matrix} r & \quad n-r \end{matrix} \\ \begin{bmatrix} P_{aa} & P_{ab} \\ & \\ P_{ba} & P_{ba} \end{bmatrix} \end{array} \begin{matrix} r \\ \\ n-r \end{matrix}$$

Define E_a and D_b as in (33) and (34), respectively. Under these conditions, the system (1) is L_2-stable if

(i) The matrix

54
$$N_1 = \begin{bmatrix} E_a + P'_{ba}\, D_b\, P_{ba} & P'_{ba}\, D_b\, P_{bb} \\ & \\ P'_{bb}\, D_b\, P_{ba} & P'_{bb}\, D_b\, P_{bb} \end{bmatrix}$$

is positive definite.

(ii) we have

55
$$\text{rank } N_1 = \text{rank } \left[N_1 \,\middle|\, \begin{matrix} I_r \\ \hline 0_{n-r} \end{matrix} \right]$$

56
$$\text{rank } P_{ab} = n-r$$

<u>Proof</u> By Lemma (24), if (55) holds, then there exists a an $\alpha>0$ such that the matrix

57 $$N_2 \triangleq N_1 - \begin{bmatrix} 2\alpha I_r & 0 \\ 0 & 0 \end{bmatrix}$$

is positive semidefinite. Let $\alpha>0$ be so chosen. From (29) - (30) we have, as before, that for $i=1,\ldots,k$

58 $$\begin{aligned} \langle x, G_i x\rangle_T &\geq (\varepsilon_i-\alpha)\ ||x||_T^2 + (\alpha/\zeta_i^2)\ ||G_i x||_T^2 \\ &\geq (\varepsilon_i-\alpha)\ ||x||_T^2 + (\alpha/\zeta^2)\ ||G_i x||_T^2, \quad \forall T \geq 0 \\ &\qquad \forall x \in L_{2e}^{n_i} \end{aligned}$$

where

59 $$\zeta = \max_i \zeta_i$$

Hence we define, as in the proof of Theorem (28), $\Lambda = I_m$, $S = I_n$, and

60 $$R = -\begin{bmatrix} E_a - \alpha I_r & 0 \\ 0 & 0 \end{bmatrix}$$

61 $$Q = -\begin{bmatrix} (\alpha/\gamma^2) I_r & 0 \\ 0 & 0 \end{bmatrix}$$

If we apply Corollary (7.2.26), the test matrix N_o becomes

$$62 \quad N_o = -(P'QP+R) = \begin{bmatrix} E_a-\alpha I_r & 0 \\ 0 & 0 \end{bmatrix} + \begin{bmatrix} P'_{ba} D_b P_{ba} & P'_{ba} D_b P_{bb} \\ P'_{bb} D_b P_{ba} & P'_{bb} D_b P_{bb} \end{bmatrix}$$

$$+ \frac{\alpha}{\zeta^2} \begin{bmatrix} P'_{aa} P_{aa} & P'_{aa} P_{ab} \\ P'_{ab} P_{aa} & P_{ab} P_{ab} \end{bmatrix}$$

$$= N_2 + \underbrace{\frac{\alpha}{\zeta^2} \begin{bmatrix} P'_{aa} P_{aa} & P'_{aa} P_{ab} \\ P'_{ab} P_{aa} & P'_{ab} P_{ab} \end{bmatrix}}_{N_3} + \begin{bmatrix} \alpha I_r & 0 \\ 0 & 0 \end{bmatrix}$$

Now, N_3 is clearly positive semidefinite. Also, the lowest $(n-r)\times(n-r)$ submatrix of N_3 is positive definite, because $P'_{ab} P_{ab}$ is positive definite (recall that $P_{ab} \in R^{r\times(n-r)}$ has rank n-r). Hence by Lemma (21), N_o is positive definite, whence by Corollary (7.2.26), the system (1) is L_2-stable. □

Theorem (52) has a narrower scope of application than Theorem (28), because (i) the interconnection matrix H is required to be non-singular, and (ii) the requirement that $P_{ab} \in R^{r\times(n-r)}$ have rank n-r automatically implies that $r \geq n-r$, i.e. $r \geq n/2$; in other words, at least half of the subsystems must be pseudo-strictly passive. However, Theorem (52) has an instability counterpart, which Theorem (28) does not; moreover, Theorem (52) also contains the single-loop result Theorem (7.1.1) as a special case, as does Theorem (28).

To prove the latter claim, consider the single-loop feedback system of (6.1.1), and suppose G_1, G_2 satisfy (47) - (49). In this case, the interconnection matrix H given by (50) is nonsingular, and

63
$$P = \begin{bmatrix} 0_\nu & I_\nu \\ -I_\nu & 0_\nu \end{bmatrix}$$

The test matrix N_1 of (54) is given by

64
$$N_1 = \begin{bmatrix} (\varepsilon+\delta) I_\nu & 0 \\ \cdot & \\ 0 & 0 \end{bmatrix}$$

It is easy to verify that all conditions of Theorem (52) are satisfied if $\varepsilon+\delta > 0$. Thus Theorem (52) contains Theorem (7.1.2) as a special case.

Next, we further specialized the results of Theorem (28), and in the process, obtain some generalizations of Corollary (7.1.7). Though the results that follows are quite conservative, they have the advantage that they involve only the interconnection matrix H , and are therefore quite easy to apply.

65 Corollary Consider the system (1). Suppose that for some integer $k \leq m$, there exist positive constants $\varepsilon_1,\ldots,\varepsilon_k$ and $\zeta_1,\ldots,\zeta_k$ such that

66
$$\langle x, G_i x\rangle_T \geq \varepsilon_i ||x||_T^2 , \quad \forall T \geq 0 , \quad \forall x \in L_{2e}^{n_i}$$

67
$$||G_i x||_T \leq \zeta_i ||x||_T , \quad \forall T \geq 0 , \quad \forall x \in L_{2e}^{n_i} \qquad \Bigg\} \; i=1,\ldots,k$$

Suppose $G_{k+1},\ldots,G_m$ satisfy

68
$$\langle x, G_i x\rangle_T \geq 0 , \quad \forall T \geq 0 , \forall x \in L_{2e}^{n_i} , \quad i=k+1,\ldots,m$$

Let $r = \sum_{i=1}^{k} n_i$ and partition the interconnection matrix H as follows:

69 $$H = \begin{array}{c} \\ \\ \\ \end{array}\begin{array}{c} \begin{array}{cc} \quad r \quad & \quad n-r \quad \end{array} \\ \begin{bmatrix} H_{aa} & H_{ab} \\ & \\ H_{ab} & H_{bb} \end{bmatrix} \end{array} \begin{array}{l} \\ r \\ \\ n-r \end{array}$$

Under these conditions, the system (1) is L_2-stable if

(i) H is positive, and

(ii) rank $H_{ab} = n-r$

Proof Apply Theorem (28) with $D_b = 0$. Then we have

70 $$M_1 = \begin{bmatrix} H'_{aa} \\ \\ H'_{ab} \end{bmatrix} E_a \, [H_{aa} \quad H_{ab}]$$

Since E_a is positive definite, it is clear from (54) that M_1 is positive semidefinite. Also, (36) holds, by inspection. Finally, to verify (ii) of Theorem (28), observe that since rank H_{ab} is n-r and E_a is positive definite, we have that $H'_{ab} E_a H_{ab}$ is positive definite. □

Remarks 1) Corollary (65) contains Corollary (7.1.7) special case. To see this, observe that the interconnection matrix for the single-loop system (7.1.1) is

71 $$H = \begin{bmatrix} 0_\nu & I_\nu \\ & \\ -I_\nu & 0_\nu \end{bmatrix}$$

which satisfies all the hypotheses of Corollary (65).

2) Corollary (65) provides an answer to the following questions: Under what conditions does an interconnection of some passive subsystems and some strictly passive subsystems with finite gain result in an overall system that is L_2-stable? The

answer is that the overall system is L_2-stable provided (i) the interconnection matrix is positive, and (ii) the interconnections from the passive subsystems to the strongly passive subsystems have the following property: If we know the interconnecting signals from the passive subsystems to the strongly passive subsystems, then we can uniquely determine the outputs of the passive subsystems. Roughly speaking, condition (ii) means that any erratic behaviour at the outputs of the passive subsystems can be detected through the interconnection signals at the strongly passive subsystems.

3) Since $H_{ab} \in R^{r\times(n-r)}$, the requirement that H_{ab} have rank $n-r$ implies that $r \geq n-r$, i.e. that $r \geq n/2$. Hence, in order to apply Corollary (65), at least half the subsystems must be strongly passive with finite gain. Theorem (72) below removes this restriction.

72 Theorem Suppose all the hypotheses of Corollary (65) hold, except that (ii) is replaced by the following condition:

(ii') whenever $v \in R^{(n-r)}$ is a nonzero solution of $H_{ab}\, v = 0$, we have

73 $$v'\, H_{bb}\, v > 0$$

Under these conditions, the system (1) is L_2-stable.

Proof Since (66) and (67) hold, choose $\alpha > 0$ such that $\alpha < \varepsilon_i$ for $i=1,\ldots,k$, and let $\zeta = \max\{\zeta_1,\ldots,\zeta_k\}$. Then, using the familiar argument, we have

74 $$\langle x, G_1 x\rangle_T \geq (\varepsilon_i-\alpha)\, ||x||_T^2 + (\alpha/\zeta^2)\, ||G_i x||_T^2 ,$$

$$\forall T \geq 0 ,\quad \forall x \in L_{2e}^{n_i} ,\quad i=1,\ldots,k$$

Now apply Theorem (7.1.18), with $\Lambda = I_n$, $S = I_n$

75 $$Q = -\begin{bmatrix} (\alpha/\zeta^2)\ I_r & 0 \\ 0 & 0 \end{bmatrix}, \quad R = -\begin{bmatrix} E_a - \alpha I_r & 0 \\ 0 & 0 \end{bmatrix}$$

where E_a is defined in (33). Then the test matrix M defined in (7.1.19) becomes

76 $$M = \begin{bmatrix} (\alpha/\zeta^2)\ I_r & 0 \\ 0 & 0 \end{bmatrix} + \begin{bmatrix} H'_{aa}\ (E_a - \alpha I_r) H_{aa} & H'_{aa}\ (E_a - \alpha I_r) H_{ab} \\ H'_{ab}\ (E_a - \alpha I_r) H_{aa} & H'_{ab}\ (E_a - \alpha I_r) H_{ab} \end{bmatrix} + (H + H')/2$$

Let R denote the sum of the last two matrices. Since R is positive semidefinite, we have by Lemma (21) that M is positive definite if the bottom $(n-r)\times(n-r)$ submatrix of R, i.e.

77 $$R_{bb} \triangleq H'_{ab}\ (E_a - \alpha I_r)\ H_{ab} + (H_{bb} + H'_{bb})/2$$

is positive definite. Since R_{bb} is positive semi-definite, we have $v'\ R_{bb}\ v \geq 0 \quad \forall v \in R^{n-r}$. So, in order to prove that R_{bb} is positive definite, it is enough to show that $v'\ R_{bb}\ v = 0$ implies $v = 0$. Accordingly, suppose $v'\ R_{bb}\ v = 0$. Then from (77) we get

78 $$v'\ R_{bb}\ v = (H_{ab}\ v)'\ (E_a - \alpha I_r)\ (H_{ab}\ v) + v'\ H_{bb}\ v = 0$$

Since $E_a - \alpha I_r$ is positive definite and H_{bb} is positive, (78) implies that

79 $$H_{ab}\ v = 0$$

80 $$v'\ H_{bb}\ v = 0$$

Now by condition (ii'), (79) and (80) together imply that $v = 0$. Hence R_{bb} is positive definite, M is positive definite by Lemma (21), and the system at hand is L_2-stable by Theorem (7.1.18). □

81 Example Consider a system of the form (1), with $m = 3$, and $n_i = 1$ for $i=1,2,3$. Suppose G_1 satisfies (66) and (67), and suppose G_2, G_3 satisfy (68). Suppose

82 $$H = \begin{bmatrix} 1 & 0 & -1 \\ 2 & 1 & 2 \\ -1 & 0 & 1 \end{bmatrix}$$

Then H is positive, but Corollary (65) can not be applied, because $r < n-r$, so that H_{ab} can not possibly have rank $n-r$. However, Theorem (72) has no such restriction. Applying Theorem (72) to the system at hand, we note first of all that H is positive. Next, it is easy to verify that condition (ii') is satisfied. Thus the given system is L_2-stable.

NOTES AND REFERENCES

The passivity theorem for feedback systems was given by Sandberg [San. 2] and Zames [Zam. 3]. The criterion allowing a "trade-off" between the forward and feedback subsystems is due to Cho and Narendra [Cho 1 and 2]. The general dissipativity criteria are due to Moylan and Hill [Moy. 2]; these generalize earlier work in [Sun. 1] and [Vid. 3]. The specializations of the general results can be found in [Vid. 8]. Another extension of the passivity theorem to large-scale systems is given by Sandberg [San. 4]. As yet, there are no satisfactory generalizations of the "multiplier" methods [Zam. 4] to large-scale systems.

CHAPTER 8: L_2-INSTABILITY CRITERIA

In this Chapter, we present several criteria for a large-scale interconnected system to be L_2-unstable. These criteria contain the instability counterparts of both the "small gain" type stability criteria of Chapter 6 and the "dissipativity" type stability criteria of Chapter 7. Many of the results here are based on an orthogonal decomposition of the input space. In Chapter 9, we show how these results can be extended to L_∞-instability using the technique of exponential weighting.

We begin by discussing the single-loop case in Section 8.1. In Section 8.2, we present instability criteria of the "small gain" type, which are the instability counterparts of the results in Chapter 6. Finally, in Section 8.3, we present instability criteria of the "dissipativity" type, which are the instability counterparts of the results of Chapter 7. These yield both "small gain" and "passivity" type instability criteria as special cases. Throughout, we use $\|.\|$ and $\|.\|_T$ to denote $\|.\|_2$ and $\|.\|_{T2}$, respectively, because we deal only with L_2-spaces.

8.1 SINGLE-LOOP SYSTEMS

In this section, we present the basic results concerning the L_2-instability of single-loop feedback systems.

To facilitate the discussion, we restate here some definitions and facts from Section 3.3.

1 <u>Definition</u> An operator $G: L_{2e}^{\nu} \to L_{2e}^{\nu}$ is said to belong to <u>class U</u> if

(i) G is linear.

(ii) The set $M(G) \subset L_2^{\nu}$ defined by

2 $$M(G) = \{x \in L_2^{\nu} : Gx \in L_2^{\nu}\}$$

is a proper subset of L_2^{ν}.

(iii) There exists a finite constant $\zeta_c(G)$ such that*

3 $$\|Gx\| \le \zeta_c(G)\ \|x\|, \quad \forall x \in M(G)$$

(iv) There exists a family of finite constants $(\alpha_T, T\in R_+)$ such that

4 $$\|(Gx)\|_T \le \alpha_T\ \|x\|_T, \ \forall x \in L_{2e}^{\nu}$$

A useful property of class U operators is brought out in the following lemma (see Lemma (3.3.12) for the proof).

5 Lemma Let $G: L_{2e}^{\nu} \to L_{2e}^{\nu}$ belong to class U. Then $M(G)$ is a closed subspace of L_2^{ν}.

As stated in Lemma (5), an operator G of class U has the property that its "set of stabilizing inputs" M(G) is a proper closed subspace of L_2^{ν}. As a result, its orthogonal complement $M^{\perp}(G)$ defined by

6 $$M^{\perp}(G) = \{z\in L_2^{\nu}: \langle z,x\rangle = 0 \quad \forall x\in M(G)\}$$

contains at least one nonzero element.

Recall that, if G is a linear convolution operator with transfer function matrix $\hat{G}(\cdot)$, and if $\hat{G}(\cdot)$ has a right-coprime factorization $(\hat{N}(\cdot), \hat{D}(\cdot))$ in $\hat{A}^{n\times n}$, then it is possible to give an explicit characterization of the set M(G) (Theorem (3.3.32). Further, if $\hat{G}(\cdot)$ has a pole in the open right half-plane, then one can demonstrate some elements of $M^{\perp}(G)$ (Theorem 3.3.45).

Theorem (7) below is the basic "small gain" type L_2-instability theorem for single-loop systems.

7 Theorem Consider a system described by

---

* Throughout this chapter, we use M(G) and $\zeta_c(G)$ instead of $M_2(G)$ and $\zeta_{c2}(G)$ in the interests of brevity.

8a $$e_1 = u_1 - y_2$$

8b $$e_2 = u_2 + y_1$$

8c $$y_1 = G_1 e_1$$

8d $$y_2 = G_2 e_2$$

where $u_1, u_2, e_1, e_2, y_1, y_2 \in L_{2e}^{\nu}$ for some positive integer ν, and G_1, G_2 map L_{2e}^{ν} into itself. Suppose G_1 belongs to Class U, and $\zeta(G_2) < \infty$. Suppose, that corresponding to each u_1, u_2 in L_2^{ν}, (8) has at least one solution in L_{2e} for e_1, e_2, y_1, y_2. Under these conditions, the system (8) is L_2-unstable if

9 $$\zeta(G_2)\ \zeta_c(G_1) \leq 1$$

In particular, we have that $y_1 \notin L_2^{\nu}$ whenever $u_2 = 0$, and $u_1 \in M^{\perp}(G_1)/\{0\}$.

Note that, in Theorem (7), instability results whenever u_1 is a nonzero element of the orthogonal complement $M^{\perp}(G_1)$. It is shown in Chapter 9 that such results can be readily adapted to encompass L_∞-instability.

Note that Theorem (7) is the stability counterpart of Theorem (6.1.1). Roughly speaking, Theorem (7) states that, if we place a stable feedback system around an unstable forward system, then the overall system is itself unstable, provided the product of the conditional gain of the unstable system and the gain of the stable system does not exceed one.

Theorem (10) is the basic passivity-type instability theorem for single-loop feedback systems.

10 <u>Theorem</u> Consider the system (8), and suppose G_1 belongs to Class U. Suppose that, for each $u_1, u_2 \in L_2^{\nu}$, (8) has

at least one solution in L_2^ν for e_1, e_2, y_1, y_2. Suppose in addition that

(i) There exists a constant ε such that

11 $$\langle x, G_1, x\rangle \geq \varepsilon \, \|x\|^2 , \quad \forall x \in M(G_1)$$

(ii) There exists a constant δ such that

12 $$\langle x, G_2 x\rangle \geq \delta \, \|G_2 x\|^2, \quad \forall x \in L_2^\nu$$

(iii) We have

13 $$G_2 x = 0 \quad \Rightarrow x = 0$$

Under these conditions, the system (8) is L_2-unstable if

14 $$\varepsilon + \delta > 0$$

Specifically, if $u_1 = 0$ and $u_2 \in M^\perp(G_1)/\{0\}$, we have that either y_1 or y_2 does not belong to L_2^ν.

An interesting feature of Theorem (10) is that the nonzero input is applied, not to the unstable system G_1, but to the possibly stable system G_2. It is still an open question whether the system (8) can be made L_2-unstable with u_2 constrained to be zero, if (14) holds.

In Theorem (10), we show that a particular choice of inputs, either y_1 or y_2 does not belong to L_2^ν. By adding an extra assumption, we can show that $y_1 \notin L_2^\nu$ for a particular choice of inputs.

15 Corollary Suppose all the hypotheses of Theorem (10) hold, and that in addition, G_2 maps L_2^ν into itself. Under these conditions, we have that $y_1 \notin L_2^\nu$ whenever $u_1 = 0$, $u_2 \in M^\perp(G_1)/\{0\}$.

8.2 CRITERIA OF THE SMALL GAIN TYPE

In this section, we present several L_2-instability criteria of the "small gain" type for large-scale interconnected systems. These results are the instability counterparts of those of Chapter 6. Note that all of the criteria given here are based on selecting some nonzero elements from an orthogonal complement $M^{\perp}(G_i)$, and can therefore be extended to L_∞-instability (see Chapter 9).

Throughout this section, we consider systems described by

1a $$e_i = u_i - \sum_{j=1}^{m} H_{ij}\, y_j \qquad i = 1,\ldots,m$$

1b $$y_i = G_i\, e_i$$

where $u_i, e_i, y_i \in L_{2e}^{n_i}$ for some positive integer n_i, $G_i : L_{2e}^{n_i} \to L_{2e}^{n_i}$, and $H_{ij} : L_{2e}^{n_j} \to L_{2e}^{n_i}$. Without stating it explicity every time, we assume that, corresponding to every set of inputs $u_i \in L_2^{n_i}$ $\forall i$, the system equations (1) have at least one solution for e_i, y_i in $L_{2e}^{n_i}$. This is a weaker assumption than requiring the system (1) to be well-posed, and is certainly satisfied if the system (1) is well-posed. Recall that some sufficient conditions for the well-posedness of systems of the form (1) are given in Chapter 5.

The first result is an instability counterpart of Theorem (6.2.50).

2 <u>Theorem</u> Consider the system (1), and suppose (i) G_i belongs to class U for all i, and (ii) H_{ij} is unbiased (i.e. $H_{ij}(0) = 0$) $\forall i,j$. Suppose there exist (not necessarily non-negative) constants α_{ij} such that

3 $$\langle e_i, -H_{ij} G_j e_j \rangle \le \alpha_{ij} \, \|e_i\| \, \|e_j\| \;\; \forall e_i \in L_2^{n_i},\; \forall e_j \in M(G_j)$$

Under these conditions, the system (1) is L_2-unstable if one can find positive constants $\lambda_1,\ldots,\lambda_m$ such that the matrix $S \in R^{m\times m}$ defined by

4 $$s_{ij} = \lambda_i \delta_{ij} - (\lambda_i \alpha_{ij} + \lambda_j \alpha_{ji})/2, \quad \delta_{ij} = \text{Kronecker delta}$$

is positive definite. In particular, whenever $u_i \in M^{\perp}(G_i)\ \forall i$ and $u_i \neq 0$ for some i, we have that either $e_i \notin L_2^{n_i}$ or $y_i \notin L_2^{n_i}$ for some i.

<u>Proof</u> Let $u_i \in M^{\perp}(G_i)\ \forall i$, and suppose by way of contradiction that $e_i \in L_2^{n_i}$, $y_i \in L_2^{n_i}\ \forall i$. Since G_i belongs to class U for all i, this implies that $e_i \in M(G_i)\ \forall i$. Now, from the system equations (1), we get

5 $$e_i = u_i - \sum_{j=1}^{m} H_{ij}\, G_j\, e_j$$

6 $$\langle e_i, e_i \rangle = \langle e_i, u_i \rangle - \sum_{j=1}^{m} \langle e_i,\, H_{ij}\, G_j\, e_j \rangle$$

$$= - \sum_{j=1}^{m} \langle e_i,\, H_{ij}\, G_j\, e_j \rangle \le \sum_{j=1}^{m} \alpha_{ij}\, \|e_i\|\, \|e_j\|$$

where we use (3) and the fact that $\langle e_i, u_i \rangle = 0$ because $u_i \in M^{\perp}(G_i)$. Next, let $\lambda_1,\ldots,\lambda_m$ be chosen such that S is positive definite. Then (6) implies that

7 $$\sum_{i=1}^{m} \lambda_i\, \|e_i\|^2 - \sum_{i=1}^{m} \sum_{j=1}^{m} \lambda_i\, \alpha_{ij}\, \|e_i\|\, \|e_j\| \le 0$$

8 $$\sum_{i=1}^{m} \sum_{j=1}^{m} s_{ij}\, \|e_i\|\, \|e_j\| \le 0$$

However, since S is positive definite, (8) implies that $e_i = 0\ \forall i$. Since G_i (being in class U) is linear, we have $G_i\, e_i = 0\ \forall i$, and since H_{ij} is unbiased, we have $H_{ij}\, G_j\, e_j = 0$ $\forall i$. This leads to a contradiction, since this implies

9 $$u_i = e_i + \sum_{j=1}^{m} H_{ij} G_j e_j = 0, \forall i$$

whereas u_i was chosen to be nonzero for at least some i. This contradiction shows that our original assumption is false, i.e. either $e_i \notin L_2^{n_i}$ or $y_i \notin L_2^{n_i}$ for some i . □

10 Corollary Suppose that the hypotheses of Theorem (2) hold, and in addition, H_{ij} maps $L_2^{n_i}$ into $L_2^{n_i}$, $\forall i,j$. Under these conditions, the system (2) is L_2-unstable. In particular, whenever $u_i \in M^{\perp}(G_i)$ $\forall i$ and $u_i \neq 0$ for some i, we have that $y_i \notin L_2^{n_i}$ for some i.

Proof Let $u_i \in M^{\perp}(G_i)$ $\forall i$, and suppose by way of contradiction that $y_i \in L_2^{n_i}$ $\forall i$. Then, since $H_{ij}(L_2^{n_j}) \subset L_2^{n_i}$ $\forall i,j$, we have from (2a) that $e_i \in L_2^{n_i}$ $\forall i$. This now leads to a contradiction, just as in the proof of Theorem (1). Hence our original assumption is false, i.e. $y_i \notin L_2^{n_i}$ for some i. □

The difference between Theorem (2) and Corollary (10) is that in Theorem (2), we show that certain inputs either produce errors that are not in $L_2^{n_i}$ or produce outputs that are not in $L_2^{n_i}$. In Corollary (10), by adding the condition that $H_{ij}(L_2^{n_j}) \subset L_2^{n_i}$ $\forall i,j$, we draw a better conclusion: namely that certain inputs produce outputs that are not in $L_2^{n_i}$.

The next result is an instability counterpart of Corollary (6.2.43).

11 Theorem Consider the system (1), and suppose G_i belongs to class U for all i. Suppose in addition that $\zeta(H_{ij}) < \infty$ $\forall i,j$. Under these conditions, the system (1) is L_2-unstable if the test matrix $P \in R^{m \times m}$ defined by

12 $$P_{ij} = \zeta(H_{ij})\ \zeta_c(G_j)$$

has a spectral radius less than or equal to one. In particular, whenever $u_i \in M^{\perp}(G_i)/\{0\}$ $\forall i$, we have that $y_i \notin L_2^{n_i}$ for some i.

<u>Proof</u> Let $u_i \in M^{\perp}(G_i)/ 0$ $\forall i$, and suppose by way of contradiction that $y_i \in L_2^{n_i}$ $\forall i$. Since $\zeta(H_{ij}) < \infty$ $\forall i,j$, this implies by (1a) that $e_i \in L_2^{n_i}$ $\forall i$. Next, $e_i \in L_2^{n_i}$, $y_i \in L_2^{n_i}$ implies that $e_i \in M(G_i)$ $\forall i$. Now define

13 $$z_i = u_i - e_i = \sum_{j=1}^{m} H_{ij} y_j$$

Since $u_i \in M^{\perp}(G_i)$ and $e_i \in M(G_i)$, we have

14 $$\|z_i\|^2 = \|u_i\|^2 + \|e_i\|^2 > \|e_i\|^2 \ \forall i \text{ since } u_i \neq 0$$

On the other hand, since

15 $$z_i = \sum_{j=1}^{m} H_{ij} y_j = \sum_{j=1}^{m} H_{ij} G_j e_j$$

we also get

16 $$\|z_i\| \leq \sum_{j=1}^{m} \|H_{ij} G_j e_j\| \leq \sum_{j=1}^{m} \zeta(H_{ij}) \zeta_c(G_j) \|e_j\|$$

$$= \sum_{j=1}^{m} p_{ij} \|e_j\|$$

Let $\rho(P)$ denote the spectral radius of P; by assumption, $\rho(P) \leq 1$. Now, since P has all nonnegative elements, the Perron-Frobenius theorem [Gan. 1, p. 66] states that $\rho(P)$ is an eigenvalue of P, and the corresponding row eigenvector v can be choosen to have all nonnegative elements (and of course, at least one positive element). Let v be so chosen. Then from (14) we get

17 $$\sum_{i=1}^{m} v_i \|z_i\| > \sum_{i=1}^{m} v_i \|e_i\|$$

On the other hand, from (16) we get

18 $$\sum_{i=1}^{m} v_i \|z_i\| \le \sum_{i=1}^{m}\sum_{j=1}^{m} v_i p_{ij} \|e_j\| = \rho(P) \sum_{j=1}^{m} v_j \|e_j\|$$

where we use the fact that v is a row eigenvector of P. Now, since $\rho(P) \le 1$, (17) and (18) contradict each other. Hence our original assumption is false, i.e. $y_i \notin L_2^{n_i}$ for some i. □

The following result is a corollary of both Theorems (2) and (11).

19 <u>Theorem</u> Consider the system (1), and suppose G_i belongs to class U for all i. Suppose in addition that $\zeta(H_{ij}) < \infty$ ∀i,j. Under these conditions, the system (1) is L_2-unstable if $\rho(P) < 1$, where $P \in R^{m\times m}$ is defined by (12). In particular, whenever $u_i \in M^{\perp}(G_i)$ ∀i and $u_i \neq 0$ for some i, we have that $y_i \notin L_2^{n_i}$ for some i.

<u>Proof</u> Let $u_i \in M^{\perp}(G_i)$ ∀i, and suppose by way of contradiction that $y_i \in L_2^{n_i}$ ∀i. Then (1a) implies that $e_i \in L_2^{n_i}$ ∀i; also, $e_i \in L_2^{n_i}$, $y_i \in L_2^{n_i}$ implies that $e_i \in M(G_i)$. Now define z_i as in (13). Since $u_i \in M^{\perp}(G_i)$ and $e_i \in M(G_i)$, we have

20 $$\|z_i\| \ge \|e_i\| \; \forall i, \quad \|z_i\| > \|e_i\| \text{ for some } i$$

because $u_i \neq 0$ for some i. Also, as in (16), we get

21 $$\|z_i\| \le \sum_{j=1}^{m} p_{ij} \|e_j\|$$

By assumption, $\rho(P) < 1$, and P has all nonnegative elements. Also note that the gains $\zeta(H_{ij})$ and $\zeta_c(G_i)$ are upper bounds, and that they can therefore be increased. So it is possible to increase the gains $\zeta(H_{ij})$ and $\zeta_c(G_i)$ such that (i) all elements of P are positive, and (ii) $\rho(P) < 1$. By the Perron-Frobenius theorem [Gan. 1, p. 66] $\rho(P)$ is an eigenvalue of P, and the corresponding row eigenvector v can be chosen such that v has all positive elements. Let v be so chosen. Then from (14) we get

22 $$\sum_{i=1}^{m} v_i \|z_i\| > \sum_{i=1}^{m} v_i \|e_i\|$$

because of (20) and the fact that $v_i > 0 \; \forall i$. On the other hand, just as in (18) we get

23 $$\sum_{i=1}^{m} v_i \|z_i\| \leq \rho(P) \sum_{j=1}^{m} v_j \|e_j\|$$

which contradicts (22). Hence our original assumption is false, i.e. $y_i \notin L_2^{n_i}$ for some i. □

As mentioned previously, Theorem (19) is a corollary of Theorem (2) as well as Theorem (11). To see that Theorem (19) is a corollary of Theorem (2), observe that if $\zeta(H_{ij}) < \infty \; \forall i,j$, then

24 $$\langle e_i, -H_{ij}G_j e_j \rangle \leq \zeta(H_{ij}) \, \zeta_c(G_j) \, \|e_i\| \, \|e_j\|$$

so that (3) is satisfied with

25 $$\alpha_{ij} = \zeta(H_{ij}) \, \zeta_c(G_j) = p_{ij}$$

Thus, if $\rho(P) < 1$, then the leading principal minors of $I_m - P$ are all positive (by Lemma (6.2.8)); therefore by Fact (6.2.11) there exist positive constants $\lambda_1, \ldots, \lambda_m$ such that $\Lambda(I_m - P) + (I_m - P)'\Lambda$ is positive definite, where

26 $$\Lambda = \text{Diag}\,\{\lambda_1, \ldots, \lambda_m\}$$

This is exactly what is required for Theorem (2) to apply. Thus Theorem (19) is a corollary of Theorem (2).

Observe also that if $\rho(P) < 1$ then $\rho(P) \leq 1$, so that the hypothesis of Theorem (11) is satisfied. In this sense,

Theorem (19) is a corollary of Theorem (11). However, the set of inputs that produce "unstable" outputs is slightly different in the two theorems. In Theorem (11), it is shown that if $\rho(P) \le 1$, then instability results if $u_i \in M^{\perp}(G_i)/\{0\}\ \forall i$. In Theorem (19), it is shown that if $\rho(P) < 1$ (which is a stronger condition), then instability is produced by a broader class of inputs, namely $u_i \in M^{\perp}(G_i)\ \forall i$ and $u_i \neq 0$ for at least one i; in other words, instability can be excited from anyone of the inputs.

Up to now, we have only considered an interconnection of several subsystems of class U. We now study interconnections of some systems of class U and other systems having finite gain.

27 Theorem Consider the system (1), and suppose that for each i, either $\zeta(G_i) < \infty$ or G_i belongs to class U, and that at least one of the G_i's belongs to class U. Also, suppose $\zeta(H_{ij}) < \infty$ for all i,j. Define the test matrix $P \in R^{m \times m}$ by

28 $$P_{ij} = \zeta(H_{ij})\ \zeta_c(G_j),\ \forall i,j$$

under these conditions, the system (1) is L_2-unstable if $\rho(P) < 1$. In particular, whenever $u_i \in M^{\perp}(G_i)$ for all i and $u_i \neq 0$ for some i, we have that $y_i \notin L_2^{n_i}$ for some i.

Remarks Note that, if $\zeta(G_i) < \infty$ for a certain i, then $M(G_i) = L_2^{n_i}$ and $\zeta_c(G_i) = \zeta(G_i)$. Also, for this value of i, $M^{\perp}(G_i) = \{0\}$, so that $u_i \in M^{\perp}(G_i)$ implies $u_i = 0$ whenever $\zeta(G_i) < \infty$. Hence the set of "destabilizing" inputs demonstrated in Theorem (27) consists of: $u_i = 0$ if $\zeta(G_i) < \infty$, $u_i \in M^{\perp}(G_i)$ if G_i belongs to class U, and $u_i \neq 0$ for some i. In particular, instability can be excited from anyone of the inputs to the class U subsystems.

Proof Since $\zeta(H_{ij})$, $\zeta_c(G_j)$ are all upper bounds, they can be increased in such a way that $P_{ij} > 0\ \forall i,j$ and $\rho(P) < 1$. Let the inputs u_i be selected as above, and suppose by way of contradiction that $y_i \in L_2^{n_i}\ \forall i$. Then (1a) implies that

$e_i \in L_2^{n_i}$ $\forall i$, which together with $y_i \in L_2^{n_i}$ implies that $e_i \in M(G_i)$ whenever G_i belongs to class U. As before, define

29 $$z_i = \sum_{j=1}^{m} H_{ij}\, y_i = u_i - e_i, \quad i = 1,\ldots,m$$

Now, if $\zeta(G_i) < \infty$, then $u_i = 0$ so that $\|z_i\| = \|e_i\|$; if G_i belongs to class U, then $\|z_i\|^2 = \|u_i\|^2 + \|e_i\|^2$ since $u_i \in M^{\perp}(G_i)$ and $e_i \in M(G_i)$. Hence

30 $$\|z_i\| \geq \|e_i\| \ \forall i$$

Also, since $u_i \neq 0$ for some i, strict inequality holds in (30) for some i. Now let $v \in R^m$ be a row eigenvector of P corresponding to the eigenvalue $\rho(P)$, chosen in such a way that $v_i > 0$ $\forall i$. Then on the one hand, we have from (30) that

31 $$\sum_{i=1}^{m} v_i \|z_i\| > \sum_{i=1}^{m} v_i \|e_i\|$$

On the other hand,

32 $$\sum_{i=1}^{m} v_i \|z_i\| \leq \sum_{i=1}^{m} v_i P_{ij} \|e_j\| = \rho(P) \sum_{i=1}^{m} v_j \|e_j\|$$

Hence (31) and (32) contradict each other, which shows that our original assumption is false; i.e., $y_i \notin L_2^{n_i}$ for some i. □

8.3 DISSIPATIVITY-TYPE CRITERIA

In this section, we derive several dissipativity type criteria for L_2-instability. Some of these are based on the idea of conditional dissipativity, while the rest make use of an orthogonal decomposition of the input space. By specializing the general dissipating-type criteria, we can obtain passivity-type criteria. The results given here are instability counterparts of those in Chapter 7.

Throughout this section, we shall be concerned with systems described by

1a $$e_i = u_i - \sum_{i=1}^{m} H_{ij}\, y_j$$

$$i = 1,\ldots,m$$

1b $$y_i = G_i\, e_i$$

where e_i, u_i, $y_i \in L_{2e}^{n_i}$ for some integer n_i, G_i: $L_{2e}^{n_i} \to L_{2e}^{n_i}$, and H_{ij} is a constant matrix of dimension $n_i \times n_j$. Throughout, we assume that, corresponding to each set of inputs $u_i \in L_2^{n_i}$ $\forall i$, the system equations (1) have a unique solution set $e_i \in L_{2e}^{n_i}$, $y_i \in L_{2e}^{n_i}$ $\forall i$, which depends in a causal way on the input set $\{u_1,\ldots,u_m\}$. This is certainly the case if the system (1) is well-posed (see Chapter 5).

8.3.1 CONDITIONAL DISSIPATIVITY APPROACH

We begin by recalling a definition from Section 3.3.

2 Definition Suppose G: $L_{2e}^{n} \to L_{2e}^{n}$, and let $Q,R,S \in R^{n\times n}$ be given, with Q and R symmetric. Then G is said to be conditionally dissipative with respect to (Q,R,S) if

3 $$\langle Gx,QGx\rangle + \langle x,Rx\rangle + \langle Gx,Sx\rangle \geq 0, \quad \forall x \in M(G)$$

It is shown in Section 3.3 that, if G is a linear convolution operator and its transfer function matrix $\hat{G}(\cdot)$ has a right-coprime factorization, then it is possible to give a necessary and sufficient condition for G to be (Q,R,S) - conditionally dissipative (see Lemma (3.3.50)).

We lead up to the main result of the subsection through a series of lemmas.

4 Lemma Consider the system (1), described in aggregated form as

5 $$e = u - Hy, \quad y = Ge$$

where $e, u, y \in L_{2e}^n$, $G: L_{2e}^n \to L_{2e}^n$, and $H \in R^{n\ n}$. If G is (Q,R,S) conditionally dissipative, then the system (5) is $(\bar{Q},\bar{R},\bar{S})$ - conditionally dissipative, where

6a $$\bar{Q} = Q + H'RH - (SH + H'S')/2$$

6b $$\bar{R} = R$$

6c $$\bar{S} = S - 2H'R$$

Proof We wish to show that, whenever $u \in L_2^n$ and $y \in L_2^n$, we have

7 $$\langle y,\bar{Q}y\rangle + \langle u,\bar{R}y\rangle + \langle y,\bar{S}u\rangle \geq 0$$

Now, if $u \in L_2^n$ and $y \in L_2^n$, then $e \in L_2^n$, from (5). Further, since G is (Q,R,S) - conditionally dissipative, we have that, whenever $e \in L_2^n$, $y \in L_2^n$,

8 $$\langle y,Qy\rangle + \langle e,Re\rangle + \langle y,Se\rangle \geq 0$$

If we replace e by u- Hy in (8), then (8) becomes (7). The details are entirely analagous to those in the proof of Lemma (7.2.6). □

9 Lemma Consider the system (1), and suppose G is (Q,R,S) - conditionally dissipative, but the system (1) is not $(\bar{Q},\bar{R},\bar{S})$ - dissipative. Then the system (1) is L_2-unstable if $\bar{Q} \leq 0$.

Proof By assumption, (1) has a unique solution y corresponding to each u, which we can write as y = Fu where $F: L_{2e}^n \to L_{2e}^n$ is causal. Also, since G is $(\bar{Q},\bar{R},\bar{S})$ - conditionally dissipative, we have from Lemma (4) that F is $(\bar{Q},\bar{R},\bar{S})$-conditionally dissipative. By Lemma (3.3.8), if $\bar{Q}\leq 0$, and F is not $(\bar{Q},\bar{R},\bar{S})$ -dissipative, then $M(F)\neq L_2^n$, i.e., the system (1) is L_2-unstable. □

The next lemma gives a general approach to L_2-instability, based on the preceding results.

10 <u>Lemma</u> Consider the system (1). Suppose that G_i is (Q_i, R_i, S_i) - conditionally dissipative for all i. Suppose one of the G_i is not (Q_i, R_i, S_i) - dissipative, and the rest are unbiased (i.e. $G_i(0) = 0$ for all other values of i). Under these conditions, the system (1) is L_2-unstable if $\bar{Q} \le 0$, where $\bar{Q}$ is given by (6a), and

11 $$Q = \text{Diag}\,\{Q_1, \ldots, Q_m\},\; R = \text{Diag}\,\{R_1, \ldots, R_m\}$$
$$S = \text{Diag}\,\{S_1, \ldots, S_m\}$$

<u>Proof</u> Let us write $y = Fu$, where F is causal. By Lemma (4), F is $(\bar{Q}, \bar{R}, \bar{S})$ - conditionally dissipative. Also $\bar{Q} \le 0$ by hypothesis. Thus, if we can show that F is <u>not</u> $(\bar{Q}, \bar{R}, \bar{S})$- dissipative, then L_2-instability would follow from Lemma (9).

To show that F is not $(\bar{Q}, \bar{R}, \bar{S})$- dissipative, we proceed as follows: since G_i is not (Q_i, R_i, S_i) - dissipative by assumption, there exist an $x_i \in L_{2e}^{n_i}$ and a $T < \infty$ such that

12 $$\langle G_i x_i, Q_i G_i x_i \rangle_T + \langle x_i, R_i x_i \rangle_T + \langle G_i x_i, S_i x_i \rangle_T < 0$$

Let us now choose

13 $$e = [\,0,\ 0,\ \ldots,\ x_i,\ 0,\ \ldots, 0]$$

14 $$y = [\,0,\ 0,\ \ldots,\ G_i x_i,\ 0,\ \ldots, 0]$$

15 $$u = e + Hy$$

Then this set of u,e,y satisfies the system equations (1). Moreover,

16 $$\langle y,\bar{Q}y\rangle_T + \langle u,\bar{R}u\rangle_T + \langle y,\bar{S}u\rangle_T = \langle y,Qy\rangle_T + \langle e,Re\rangle_T + \langle y,Se\rangle_T$$

$$= \langle G_i x_i, QG_i x_i\rangle_T + \langle x_i, R_i x_i\rangle_T + \langle G_i x_i, S_i x_i\rangle_T$$

$$< 0$$

Hence F is not $(\bar{Q},\bar{R},\bar{S})$ - dissipative.□

Remark Instead of requiring G_i to be unbiased, we can suppose that $\langle G_i 0, Q_i G_i 0\rangle_T \le 0$ for all T. This latter condition is satisfied if $G_i 0 = 0$ or if $Q_i \le 0$.

Lemma (10) gives a general approach to deriving L_2-instability criteria, but is itself not a specific instability criterion, because it is not yet clear how to verify whether a particular operator G_i is (Q_i,R_i,S_i) - conditionally dissipative but not (Q_i,R_i,S_i) - dissipative. We now give a few specific instability criteria, based on Lemma (10). To avoid repetition, we state below two standing assumptions:

(A1) $m = 2$, and the operator G_i is (Q_i,R_i,S_i)-conditionally dissipative for $i = 1,2$

(A2) $G_2(0) = 0$, i.e., G_2 is unbiased.

The above assumptions are quite natural. In Lemma (10), it is assumed that one of the operators G_i is conditionally dissipative but not dissipative with respect to (Q_i,R_i,S_i). Without loss of generality, we can renumber the G_i's so that this operator becomes G_1; we can then "aggregate" the remaining operators into G_2, which must satisfy (A2) in order for Lemma (10) to apply.

17 Theorem Consider the system (1), and suppose (A1), (A2) hold. Suppose $N(G_1) \neq L_2^{n_1}$. Under these conditions, the system (1) is L_2-unstable if $\bar{Q} \le 0$ and $Q_1 < 0$.

Proof The theorem follows from Lemma (10) if we show that G_1 is not (Q_1,R_1,S_1) - dissipative. Since $M(G_1) \neq L_2^{n_1}$, pick $x \in L_2^{n_1}$ such that $G_1 x \in L_2^{n_1}$. Then $\|G_1 x\|_T$ becomes unbounded as $T \to \infty$. Now, since $Q_1 < 0$, it is clear that as $T \to \infty$, the quantity

18 $$\langle G_1x, QG_1x\rangle_T + \langle x, R_1x\rangle_T + \langle G_1x, S_1x\rangle_T$$

eventually becomes negative. This shows that G_1 is not (Q_1,R_1,S_1) - dissipative. □

Theorem Consider the system (1), and suppose it is well-posed. Suppose (A1) and (A2) hold, and suppose $M(G_1) \neq L_2^{n_1}$, but $\zeta_c(G_1) < \infty$. Under these conditions, the system (1) is L_2-unstable if $\bar{Q} \leq 0$, $Q_1 \leq 0$, and one of them is negative definite.

Proof If $\bar{Q} \leq 0$ and $Q_1 < 0$, the result follows from Theorem (17) above. So suppose that $\bar{Q} < 0$ and $\bar{Q}_1 \leq 0$. By assumption, G_1 is (Q_1,R_1,S_1) - conditionally dissipative, i.e.,

20 $$\langle G_1x, Q_1G_1x\rangle + \langle x, R_1x\rangle + \langle G_1x, S_1x\rangle \geq 0, \quad \forall x \in M(G_1)$$

Now, since $\zeta_c(G_1) < \infty$, we also have, for all $\alpha > 0$, that

21 $$\langle G_1x, -\alpha G_1x\rangle + \langle x, \alpha\zeta_c^2(G_1)x\rangle \geq 0, \quad \forall x \in M(G_1)$$

Adding (20) and (21) gives

22 $$\langle G_1x, (Q_1-\alpha I_{n_1})G_1x\rangle + \langle x, (R_1+\alpha\zeta_c^2(G_1)I_{n_1})x\rangle + \langle G_1x, S_1x\rangle \geq 0, \quad \forall\, x \in M(G_1)$$

Hence G_1 is also $(\tilde{Q}_1,\tilde{R}_1,\tilde{S}_1)$ - conditionally dissipative, where

23 $$\tilde{Q}_1 = Q_1-\alpha I_{n_1}, \quad \tilde{R}_1 = R_1+\alpha\zeta_c^2(G_1)I_{n_1}, \quad \tilde{S}_1 = S_1$$

However, since $\tilde{Q}_1 < 0$, G_1 is not $(\tilde{Q}_1,\tilde{R}_1,\tilde{S}_1)$ - dissipative*. Now, the original "test matrix" $\bar{Q}$ is $Q + H'RH - (SH + H'S')/2$. If we replace Q_1,R_1,S_1 by $\tilde{Q}_1,\tilde{R}_1,\tilde{S}_1$, the new "test matrix" is

24 $$\tilde{Q} = \bar{Q} - \begin{bmatrix} \alpha I_{n_1} & 0 \\ 0 & 0 \end{bmatrix} + H' \begin{bmatrix} \alpha \zeta_c^2(G_1)\ I_{n_1} & 0 \\ 0 & 0 \end{bmatrix} H$$

Since $\bar{Q} < 0$, the new test matrix $\tilde{Q} \le 0$ for sufficiently small α. Hence the system (1) is L_2-unstable, by Theorem (17). □

Both Theorems (17) and (19) permit G_1 to be nonlinear, in contrast with results that require G_1 to be class U (and hence linear). Moreover, none of the G_i need be unstable. However, since they do not construct a destabilizing input in terms of an orthogonal decomposition of the input space, they can not be extended to L_∞-instability.

8.3.2 CRITERIA BASED ON ORTHOGONAL DECOMPOSITION

In this subsection, we present some criteria for L_2-instability based on an orthogonal decomposition of the input space. These criteria contain several passivity-type instability criteria as special cases.

Throughout this subsection, we assume that the interconnection matrix H is nonsingular. To rationalize this assumption, observe that if H is nonsingular, then the following two statements are equivalent (their equivalence follows readily from (1a)):

---

* By Lemma (3.2.10), if G_1 is $(\tilde{Q}_1,\tilde{R}_1,\tilde{S}_1)$ - dissipative with $\tilde{Q}_1 < 0$, then $M(G_1) = L_2^{n_1}$. Since this is not the case, G_1 is not $(\tilde{Q}_1,\tilde{R}_1,\tilde{S}_1)$ - dissipative.

(i) $u \in L_2^n \Rightarrow e \in L_2^n$

(ii) $u \in L_2^n \Rightarrow y \in L_2^n$

Also, if H is nonsingular, the following two statements are equivalent:

(iii) The relation between u and e has finite gain

(iv) The relation between u and y has finite gain.

In other words, the nonsingularity of H is a kind of nondegeneracy assumption, which allows us to determine the stability of the system by examining the behaviour of only e, or of only y, without having to examine the other.

We now present the basic criterion.

25 Theorem Consider the system (1), and suppose (A1),(A2) hold. Suppose in addition that G_1 belongs to class U, and that the interconnection matrix H is nonsingular: Let

26
$$P = H^{-1} = \begin{array}{c} \begin{matrix} n_1 & n_2 \end{matrix} \\ \begin{bmatrix} P_{11} & P_{12} \\ P_{21} & P_{22} \end{bmatrix} \end{array} \begin{matrix} n_1 \\ n_2 \end{matrix}$$

Under these conditions, the system (1) is L_2-unstable if

(i) S_1 is nonsingular;

(ii) $Q_1 \leq 0$

(iii) The "test matrix" A defined by

27
$$A = -R - \begin{bmatrix} P_{21}' \\ P_{22}' \end{bmatrix} Q_2 [P_{21} \quad P_{22}] + \frac{P'S+S'R}{2}$$

is positive definite.

In particular, whenever

28 $$u_1 = H_{11}(S_1^{-1})'v,\ u_2 = H_{21}(s_1^{-1})'v, \text{ for some } v \in M^{\perp}(G_1)/\{0\}$$

we have that the corresponding e or y belongs to L_{2e}^n / L_2^n.

<u>Proof</u> Select u_1 and u_2 as in (28), and suppose by way of contradiction that the corresponding e,y both belong to L_2^n. Since $y_1 \in L_2^{n_1}$ and $e_1 \in L_2^{n_1}$, it follows that $e_1 \in M(G_1)$, by the definition of $M(G_1)$. Now, since

29 $$u = \begin{bmatrix} u_1 \\ u_2 \end{bmatrix} = \begin{bmatrix} H & \begin{matrix} (S_1^{-1})'v \\ 0 \end{matrix} \end{bmatrix}$$

and since $P = H^{-1}$, we have

30 $$S'Pu = \begin{bmatrix} v \\ 0 \end{bmatrix}$$

Also, from (1a), we have $y = H^{-1}(u-e) = P(u-e)$, so that

31 $$\langle y,Se\rangle = \langle e,S'y\rangle = \langle e,S'Pu\rangle - \langle e,S'Pe\rangle = -\langle e,S'Pe\rangle$$

where we use the fact that

32 $$\langle e,S'Pu\rangle = \langle e_1,v\rangle = 0 \text{ because } e_1 \in M(G_1),\ v \in M^{\perp}(G_1)$$

On the other hand, since G_i is (Q_i,R_i,S_i) - conditionally dissipative, we have

33 $$\begin{aligned} \langle y,Se\rangle &\geq -\langle e,Re\rangle - \langle y,Qy\rangle \\ &\geq \langle e,Re\rangle - \langle y_2,Q_2y_2\rangle \end{aligned}$$

since $Q_1 \leq 0$. Note now that, since $y = P(u-e)$, we have

34 $$y_2 = P_{21}(u_1-e_1) + P_{22}(u_2-e_2) = -(P_{21}e_1+P_{22}e_2)$$

because $P_{21}u_1 + P_{22}u_2 = 0$ from (29). Hence (33) becomes

35 $$\langle y,Se\rangle \geq -\langle e,Re\rangle - \langle P_{21}e_1+P_{22}e_2, Q_2(P_{21}e_1+P_{22}e_2)\rangle$$

Combining (31) and (35) gives

36 $$-\langle e,S'Pe\rangle \geq -\langle e,Re\rangle - \langle P_{21}e_1+P_{22}e_2, Q_2(P_{21}e_1+P_{22}e_2)\rangle$$

37 $$0 \geq -\langle e,Re\rangle - \langle P_{21}e_1 + P_{22}e_2, Q_2(P_{21}e_1+P_{22}e_2)\rangle + \\ + \langle e,S'Pe\rangle = \langle e,Ae\rangle$$

Since $A > 0$ by hypothesis, (37) implies that $e = 0$. In turn, this implies that $y_1 = G_1e_1 = 0$ (because G_1 is linear) and $y_2=0$ from (34). Finally, we have $S'Pu = S'P(e + Hy) = 0$. However, this contradicts (30), because v is a <u>nonzero</u> element of $M^{\perp}(G_1)$. This contradiction shows that our original assumption is false; i.e., either e or y does not belong to L_2^n. □

38 <u>Theorem</u> Theorem (25) remains valid if condition (iii) is replaced by

39 (iv) $A \geq 0$, and $\text{rank}\left[A \middle| \frac{I_{n_1}}{0}\right] = \text{rank}\, A$

40 (v) $(G_2+P_{22})e_2 = 0 \Rightarrow e_2 = 0.$

<u>Proof</u> Just as before, we arrive at (37). Now, as can be easily shown, it follows from (37) and (39) that $Ae=0$ and $e_1=0$. Hence, from (34), we get $y_2 = G_2e_2 = -P_{22}e_2$, i.e., $(G_2+P_{22})\, e_2 = 0$. Form (40), this implies that $e_2 = 0$ as well. The rest of the proof follows that of Theorem (25). □

It is of interest to compare the constructive criterion of Theorem (25) with the nonconstructive criterion of Theorem (19). In Theorem (19), we assume that $\zeta_c(G_1) < \infty$, which is also assumed in Theorem (25), because $G_1 \in$ class U implies that $\zeta_c(G_1) < \infty$. Also, G_1 must be linear in Theorem (25) but not in Theorem (19). Similarly, both Theorems (19) and (25) assume that $Q_1 \leq 0$. Thus, the only extra assumptions in Theorem (25) are that S_1 and H are nonsingular. The nonsingularity of H_1 has been discussed previously. The nonsingularity of S_1 seems to be a technical condition, whose significance is not clear. Finally, note that the hypothesis "$A > 0$" in Theorem (25) is slightly stronger than "$\bar{Q} < 0$" in Theorem (19) because, as can be easily verified,

41 $$A = -P'(\bar{Q} - \begin{bmatrix} Q_1 & 0 \\ 0 & 0 \end{bmatrix})P$$

In other words, if $A > 0$, then $\bar{Q} < 0$; also, if $Q_1 = 0$, then $A > 0$ if and only if $\bar{Q} < 0$. To summarize, it is not necessary to tack on too many unnatural conditions to turn the non-constructive criterion of Theorem (19) into the constructive criterion of Theorem (25). One only has to (i) assume that $G \in$ class U, instead of just satisfying $\zeta_c(G_1) < \infty$, (ii) strengthen "$\bar{Q} < 0$" to "$A > 0$", and (iii) assume that S_1, H are nonsingular.

By specializing Theorem (25), one can obtain various "passivity-type" instability criteria.

Theorem (42) below is the most general instability result of the passivity type, and is an instability counterpart of Theorem (7.3.52).

42 Theorem Consider the system (1), and suppose that $G_1,\ldots,G_k$ belong to class U for some integer $k \leq m$. Suppose there exist real constants $\varepsilon_1,\ldots,\varepsilon_k$ and $\delta_{k+1},\ldots,\delta_m$ such that

43 $$\langle x, G_i x\rangle \geq \varepsilon_i \; \|x\|^2 \;\forall x \in M(G_i),\; i = 1,\ldots,k$$

44 $$\langle x, G_i x\rangle \geq \delta_i \; \|G_i x\|^2,\; \forall x \in L_{2e}^{n_i},\; i = k+1,\ldots,m$$

Suppose H is nonsingular, and let $P = H^{-1}$. Under these conditions, the system (1) is L_2-unstable if

(i) the matrix

45 $$A_1 = E + P' D P$$

is positive semidefinite, where

46 $$E = \text{Diag}\,\{\varepsilon_1 I_{n_1},\ldots,\varepsilon_k I_{n_k},\ 0,\ldots,0\}$$

47 $$D = \text{Diag}\,\{0,\ldots,0,\ \delta_{k+1} I_{n_{k+1}},\ldots,\delta_m I_m\}$$

(ii) M_2 satisfies

48 $$\text{rank } M_2 = \text{rank } [M_2 | \tfrac{I_r}{0}]$$

where $r = \sum_{i=1}^{k} n_i$.

(iii) Let $P = H^{-1}$ be partitioned in the same way as H is partitioned in (41). Then

49 $$G_i e_i + \sum_{j=k+1}^{m} P_{ij} e_j = 0 \text{ for } i = k+1,\ldots,m \Rightarrow e_i = 0 \quad \text{for } i = k+1,\ldots,m$$

(iv) H is positive.

In particular, if we select inputs of the form

50 $$u_i = \sum_{j=1}^{k} H_{ij} v_j, \quad i = 1,\ldots,m$$

where $v_j \in M'(G_j)$ for $j = 1,\ldots,k$ and $v_j \neq 0$ for at least one j, then $y_j \notin L_2^{n_i}$ for some i.

<u>Proof</u> Apply Theorem (25) with

51 $$Q = -D,\ R = -E,\ S = I_n$$

Then conditions (i) and (ii) of Theorem (25) are automatically satisfied because $Q_1 = 0$, and $S_1 = I_r$. Also, the "test matrix" A becomes

52 $$A = E + P'DP + (P'+P)/2 = A_1 + (P'+P)/2$$

If P is positive, then $A \geq A_1$. However, since A_1 may not be positive definite, we have to slightly modify the proof of Theorem (25). Everything is the same until (37), which now becomes

53 $$0 \geq \langle e, Ae \rangle \geq \langle e, A_1 e \rangle$$

Now, (53) and (48) show that $e_1 = 0$. Next, $e_1 = 0$ plus (49) show that $e_2 = 0$ (just as in the proof of Corollary (38)). The rest of the proof follows that of Theorem (25). □

54 Corollary Consider the system (1), and suppose $G_1, \ldots, G_k$ belong to class U for some integer $k \leq m$. Suppose there exists positive constants $\varepsilon_1, \ldots, \varepsilon_k$ such that

55 $$\langle x, G_1 x \rangle \geq \varepsilon_i \|x\|^2, \quad \forall x \in M(G_i), \quad i = 1, \ldots, k$$

and suppose

56 $$\langle x, G_i x \rangle_T \geq 0, \quad \forall T \in L_{2e}^{n_i}, \quad i = k+1, \ldots, m$$

Under these conditions, the system (1) is L_2-unstable if (49) holds.

Proof In this case, we have $D = 0$, so that A_1 of (45) becomes

57 $$A_1 = \begin{bmatrix} E_a & 0 \\ 0 & 0 \end{bmatrix}$$

where

58 $$E_a = \text{Diag}\,\{\varepsilon_1 I_{n_1}, \ldots, \varepsilon_k I_{n_k}\}$$

Clearly A_1 is positive semidefinite and satisfies (48). x

NOTES AND REFERENCES

The orthogonal decomposition approach to the L_2-instability of feedback systems was introduced by Takeda and Bergen [Tak. 1], and extended to large-scale systems in [Vid. 2,3,7,10]. The nondissipativity approach is given by Moylan and Hill [Moy. 2] and some further results are in [Vid. 10]. Additional results on L_2-instability can be found in [Mic. 1].

CHAPTER 9: L_∞-STABILITY AND L_∞-INSTABILITY USING EXPONENTIAL WEIGHTING

In Chapters 6 and 7, we have presented two distinct approaches to obtaining stability criteria for large-scale interconnected systems, namely the "small gain" approach and the "dissipativity" approach. Certain features of these two approaches can now be stated. The small gain approach can be applied to systems set in an arbitrary space L_{pe}, for any $p \in [1,\infty]$. In contrast, the dissipativity approach requires an inner product structure on the input spaces, and can therefore be directly used only to study L_2-stability. This is a point in favour of the small gain approach. On the other hand, the small gain approach invariably leads to "weak interaction" type of results, which are sign-insensitive and are therefore conservative. In contrast, the dissipativity approach leads to results that are sign sensitive, and less conservative than those obtained by the small gain approach. Unfortunately, the dissipativity approach of Chapter 7 can not be directly applied to study L_∞-stability, because L_∞ is not an inner product space. So, what we propose in this chapter is a method of studying the L_∞-stability of a system by studying instead a property resembling the L_2-stability of an associated system. Since both dissipativity as well as small gain methods can be used to study this property of the associated system, it is clear that the results given here in effect extend the scope of applicability of the dissipativity approach to include L_∞-stability studies.

There is also another point worth mentioning. Generally speaking, the criteria for L_∞-stability obtained by directly using the small gain approach involve calculations in the time domain. The reason for this is that the L_∞-induced norm of a convolution operator is calculated in the time domain (see Lemma (3.1.51)). In contrast, the L_2-stability criteria obtained by using either the small gain or dissipativity approaches usually involve frequency domain calculations. The reason is that the L_2-induced norm of a convolution operator is calculated in the frequency domain, and the matrices Q,R, and S encountered throughout Chapter 7 are also calculated in the frequency domain (see (3.1.55) and Section 3.2). Thus, as a consequence of the results given in this chapter, one is able to obtain <u>frequency domain</u> criteria for L_∞-stability. Such results are very useful because they can be readily applied to experimental data, which are quite often obtained by means of frequency response tests.

Finally, the instability results of Chapter 8 pertain to L_2-instability, because they too make use of an inner product structure on the input space. By making appropriate use of "exponential weighting", we show in this chapter that a given system is L_∞-unstable whenever an associated system is L_2-unstable .

9.1 GENERAL STABILITY RESULT

The general result given here depends on the technique of "exponential weighting", which is described briefly.

1 <u>Definition</u> Given a real number α, the operator W_α: $L_{2e}^{n_i} \to L_{2e}^{n_i}$ is defined by

2 $$(W_\alpha x)(t) = x(t)\ \exp(\alpha t), \quad \forall x \in L_{2e}^{n_i}$$

It is clear that W_α merely "weights" each function $x(\cdot) \in L_{2e}^{n_i}$ by the exponential term $\exp(\alpha t)$. Also, for each α, $W_\alpha^{-1} = W_{-\alpha}$ maps $L_{2e}^{n_i}$ onto itself.

If α is clear from the context, we usually denote $W_\alpha x$ by x^*. Similarly, if G_i: $L_{2e}^{n_i} \to L_{2e}^{n_i}$, we employ the notation

3 $$G_i^* = W_\alpha \cdot G_i \cdot W_\alpha^{-1}$$

Note that, if $y_i = G_i\ e_i$, then

4 $$y_i^* = W_\alpha\ G_i\ e_i = W_\alpha\ G_i\ W_\alpha^{-1}\ e_i^*$$
$$= G_i^*\ e_i^*$$

In particular, suppose G_i is a convolution operator with impulse response matrix $G_i(\cdot)$; i.e., suppose

5 $$y_i(t) = \int_0^t G_i(t-\tau)\; e_i(\tau)d\tau$$

Then

6 $$y_i^*(t) = \int_0^t \exp(\alpha t)\; G_i(t-\tau)\; e_i(\tau)d\tau$$

$$= \int_0^t G_i(t-\tau)\; \exp(\alpha(t-\tau))\; .\; e_i(\tau)\; \exp(\alpha\tau)\; d\tau$$

$$= \int_0^t G_i^*(t-\tau)\; e_i^*(\tau)\; d\tau$$

where the asterix denotes exponential weighting and not matrix transposition. In other words, G_i^* is also a convolution operator whose impulse response matrix is $G_i^*(\cdot)$. Equivalently, $\hat{G}^*(j\omega) = \hat{G}(\alpha + j\omega)$.

In the same vein, suppose G_i is a memoryless nonlinearity in the sector [a,b]; i.e., suppose

7 $$y_i(t) = (G_i e_i)(t) = \phi_i(t,\; e_i(t))$$

where ϕ_i: $R_+ \times R^{n_i} \to R^{n_i}$ and

8 $$[\phi_i(t,\sigma) - a\sigma]'\; [\phi_i(t,\sigma) - b\sigma] \leq 0,\; \forall t \geq 0,\; \forall \sigma \in R^{n_i}$$

Then

9 $$y_i^*(t) = \phi_i^*(t, e_i^*(t))$$

where

10 $$\phi_i^*(t,\sigma) = \exp(\alpha t)\; \phi_i(t,\; \exp(-\alpha t)\sigma)$$

Hence

11 $$[\phi_i^*(t,\sigma) - a\sigma]'\; [\phi_i^*(t,\sigma) - b\sigma]$$

$$= [\exp \alpha t\ \phi_i(t,\exp(-\alpha t)\sigma) - a\sigma]'.$$

$$[\exp \alpha t\ \phi_i(t, \exp(-\alpha t)\sigma) - b\sigma]$$

$$= \exp(2\alpha t)[\phi_i(t,\exp(-\alpha t)\sigma) - a \exp(-\alpha t)\sigma]'.$$

$$[\phi_i(t,\exp(-\alpha t)\sigma) - b \exp(-\alpha t)\sigma]$$

$$\leq 0,\ \forall t \geq 0,\ \forall\ \sigma \in R^{n_i}$$

which shows that ϕ_i^* also belongs to the sector $[a,b]$.

In addition to the concept of exponential weighting, we also need that of "decaying L_1-memory".

12 Definition An operator $G_i: L_{2e}^{n_i} \to L_{2e}^{n_i}$ is said to have decaying L_1-memory if there exists a nonnegative-valued, nonincreasing function $m_i(\cdot) \in L_1$ such that

13 $$\|(G_i x)(t)\|^2 \leq \int_0^t m_i(t-\tau)\ \|x(\tau)\|^2\ d\tau,\ \forall t \geq 0,\ \forall x \in L_{2e}^{n_i}$$

14 Lemma Let $G_i: L_{2e}^{n_i} \to L_{2e}^{n_i}$ be a convolution operator with impulse response matrix $G_i(\cdot)$. Then the operator G_i has decaying L_1-memory if the function $t \to G_i(t) \exp(\alpha t) \in L_2^{n_i \times n_i}$ for some $\alpha > 0$.

Proof We have

15 $$(G_i x)(t) = \int_0^t G_i(t-\tau)\ x(\tau)\ d\tau$$

$$= \int_0^t G_i(t-\tau)\exp(\alpha(t-\tau))\ x(\tau)\exp(-\alpha(t-\tau))d\tau$$

Hence by Schwarz's inequality,

16 $$\|(G_i x)(t)\|^2 \leq \int_0^t \|G_i(t-\tau)\|^2\ \exp(2\alpha(t-\tau))d\tau$$

$$\cdot \int_0^t \|x(\tau)\|^2 \exp(-2\alpha(t-\tau))d\tau$$

$$= \int_0^t \|G_i(\tau)\|^2 \exp(2\alpha\tau)d\tau \cdot \int_0^t \|x(\tau)\|^2 \exp(-2\alpha(t-\tau))d\tau$$

$$= \int_0^t m_{i0} \exp(-2\alpha(t-\tau))\|x(\tau)\|^2$$

where

17 $$m_{i0} \triangleq \int_0^\infty \|G_i(\tau)\|^2 \exp(2\alpha\tau)d\tau < \infty$$

Hence (13) is satisfied with

18 $$m_i(t) = m_{i0} \exp(-2\alpha t)$$

and the proof is complete. □

19 Lemma Suppose G_i: $L_{2e}^{n_i} \to L_{2e}^{n_i}$ has decaying L_1-memory, $V \in L_\infty$, and that $e_i \in L_{2e}^{n_i}$ has the property that, for some $\alpha > 0$, we have

20 $$\|W_\alpha e_i\|_{T2} \le \|V\|_\infty \exp(\alpha T), \ \forall T \ge 0$$

Then $G_i e_i \in L_\infty^{n_i}$, and $\|G_i e_i\|_\infty$ is bounded by a constant times $\|V\|_\infty$.

Proof (20) can be rewritten as

21 $$\int_0^t \|e_i(\tau)\|^2 \exp(-2\alpha(t-\tau)) \, d\tau \le r^2, \quad \forall t$$

where we let r denote $\|V\|_\infty$ for brevity. Define

22 $$f(t) = \int_0^t \|e_i(\tau)\|^2 \exp(-2\alpha(t-\tau))d\tau$$

Then $f(t) \le r^2 \ \forall t$, by (21). Moreover, we have from (22) that

23 $$\dot{f}(t) + 2\alpha f(t) = \|e_i(t)\|^2$$

Now, since G_i has decaying L_1-memory, we have by (13) that

24 $$\|(G_i e_i)(t)\|^2 \le \int_0^t m_i(t-\tau)\ \|e_i(\tau)\|^2\, d\tau = \int_0^t m_i(t-\tau)\ [\dot{f}(\tau) + 2\alpha f(\tau)]\, d\tau$$

Using integration by parts in (24), we get

25 $$\|(G_i e_i)(t)\|^2 \le m_i(0)f(t) - m_i(t)f(0) + \int_0^t f(\tau)\dot{m}_i(t-\tau)\,d\tau + \int_0^t 2\alpha f(\tau)\ m_i(t-\tau)\ d\tau$$

However, since $m_i(\cdot)$ is nonnegative-valued and nonincreasing, (25) implies that

26 $$\|(G_i e_i)(t)\|^2 \le [m_i(0) + \int_0^t 2\alpha m_i(\tau)\ d\tau]\, r^2$$

where we also use the fact that $0 \le f(\tau) \le r^2 \forall \tau$. Finally, since $m_1(\cdot) \in L_1$, we get from (26) that

27 $$\|(g_i e_i)(t)\|^2 \le r^2 . [m_i(0) + 2\alpha\ \|m_i(\cdot)\|_1]$$

Hence $G_i e_i \in L_\infty^{n_i}$, and in fact $\|G_i e_i\|_\infty$ is bounded by a constant times r, i.e., a constant times $\|V\|_\infty$. □

<u>Remark</u> Suppose $e_i \in L_\infty^{n_i}$; then it is easy to see that, for any $\alpha > 0$, we have

28 $$\|W_\alpha e\|_{T2} \le \frac{\|e\|_\infty}{(2\alpha)^{1/2}} \exp(\alpha T), \quad \forall T \ge 0$$

Hence, if $G_i\colon L_{2e}^{n_i} \to L_{2e}^{n_i}$ has decaying L_1-memory, then $\zeta_\infty(G_i) < \infty$, in view of Lemma (19).

We now present a general result, whereby the L_∞-stability of a given large-scale interconnected system can be deduced by studying a property resembling the L_2-stability of an associated system.

29 **Theorem** Consider a system described by

30a
$$e_i = u_i - \sum_{j=1}^{m} H_{ij} y_j$$
$$i = 1,\ldots,m$$
30b
$$y_i = G_i e_i$$

where u_i, e_i, y_i all belong to $L_{2e}^{n_i}$ for some integer n_i, G_i: $L_{2e}^{n_i} \to L_{2e}^{n_i}$, and H_{ij}: $L_{2e}^{n_j} \to L_{2e}^{n_i}$. Now consider the associated system

31a
$$d_i = u_i - \sum_{j=1}^{m} H_{ij}^{*} z_i$$
$$i = 1,\ldots,m$$
31b
$$z_i = G_i^{*} d_i$$

where $G_i^{*} = W_\alpha G_i W_{-\alpha}$ and $H_{ij}^{*} = W_\alpha H_{ij} W_{-\alpha}$, as defined previously. Suppose $\zeta_\infty(H_{ij}) < \infty$ for all i,j, and suppose we can find a number $\alpha > 0$ and an integer $k \le m$ such that

(i) For $i = 1,\ldots,k$, G_i has decaying L_1-memory.

(ii) For $i = k+1,\ldots,m$, we have $\zeta_\infty(G_i)<\infty$.

(iii) $H_{ij} = 0$ for $k+1 \le i,j \le m$.

(iv) the system (31) satisfies, for some constant $\zeta_0 < \infty$, the condition

32
$$\|d_i\|_{T2} \le \zeta_0 \sum_{j=1}^{m} \|v_j\|_{T2}, \ \forall T \ge 0, \ \forall v_j \in L_{2e}^{n_i}, \quad i=1,\ldots,k$$

Under these conditions, the system (30) is L_∞-stable.

Proof Multiplying both sides of (30) by W_α, we get the following system description, which is equivalent to (30):

$$e_i^* = u_i^* - \sum_{j=1}^{m} H_{ij}^* y_j^* \qquad (33a)$$

$$y_i^* = G_i^* e_i^* \qquad (33b)$$

$$\Bigg\} \; i=1,\ldots,m$$

Now, (33) is of the form (31), with e_i^*, u_i^*, y_i^* playing the roles of d_i, v_i, and z_i, respectively. Suppose now that $u_i \in L_\infty^{n_i}$ $\forall i$. We will show that e_i, $y_i \in L_\infty^{n_i}$ $\forall i$, and that $\|e_i\|_\infty$, $\|y_i\|_\infty$ are bounded by a constant times $\sum_{i=1}^{m} \|u_i\|_\infty$. To do this, observe first of all that if $u_i \in L_\infty^{n_i}$, then $u_i^* \in L_{2e}^{n_i}$, and moreover,

34 $$\|u_i^*\|_{T2}^2 = \int_0^T \|u_i(t)\|^2 \exp(2\alpha t)\, dt$$
$$\le \|u_i\|_\infty^2 \int_0^T \exp(2\alpha t)\, dt$$
$$\le \|u_i\|_\infty^2 \exp(2\alpha T)/2\alpha$$

so that

35 $$\|u_i^*\|_{T2} \le c\, \|u_i\|_\infty \exp(\alpha T), \; \forall i$$

where $c = (1/2\alpha)^{1/2}$. Since (32) is assumed to hold, it follows that

36 $$\|e_i^*\|_{T2} \le c\, [\sum_{j=1}^{m} \|u_j\|_\infty] \exp(\alpha T), \; i=1,\ldots,k$$

Since G_i has a decaying L_1-memory for $i=1,\ldots,k$, we have from Lemma (19) that $y_i \in L_\infty^{n_i}$ for $i=1,\ldots,k$, and moreover, there exists a finite constant δ_0 such that

37 $$\|y_i\|_\infty \le \delta_0 \sum_{j=1}^{m} \|u_j\|_\infty, \quad i=1,\ldots,k$$

Since $H_{ij} = 0$ for $i,j = k+1,\ldots,m$, we have

38 $$e_i = u_i - \sum_{j=1}^{k} H_{ij}\, y_j, \quad i=k+1,\ldots,m$$

Since $\zeta_\infty(H_{ij}) < \infty$ $\forall i,j$, (38) shows that $e_i \in L_\infty^{n_i}$ for $i = k+1,\ldots,m$, and that there exists a finite constant δ_1 such that

39 $$\|e_i\|_\infty \le \delta_1 \sum_{j=1} \|u_j\|_\infty \, , \quad i=k+1,\ldots,m$$

Since $\zeta_\infty(G_i) < \infty$ $\forall i$, we next have that $y_i \in L_\infty^{n_i}$ for $i=k+1,\ldots,m$, and also the finite gain property analogous to (37). Finally, from (30a), we get

40 $$e_i = u_i - \sum_{j=1}^{m} H_{ij}\, y_j, \quad i=1,\ldots,k$$

Since $\zeta_\infty(H_{ij}) < \infty$, (40) shows that $e_i \in L_\infty^{n_i}$ for $i=1,\ldots,k$, and we also have the finite gain property analogous to (39). Hence the system (30) is L_∞-stable. □

Remarks First of all, consider the hypotheses of Theorem (29). Conditions (i) and (ii) state that all subsystem and interconnection operators have finite gain with zero bias when viewed as operators on L_∞-spaces of appropriate dimension, and that some of the subsystem operators have decaying L_1-memory. Condition (iii) states that the remaining subsystems don't directly interact among themselves. Condition (iv) is the crucial condition. It states that the associated system (31) is "L_2-stable" in a special sense, namely that it satisfies (32). It is clear from (32) that if $v_i \in L_2^{n_i}$ $\forall i$, then $d_i \in L_2^{n_i}$ for $i=1,\ldots,k$. However, in general, (32) is not the same as requiring the relation from $v_1,\ldots,v_m$ to $d_1,\ldots,d_k$ to be L_2-stable. To establish this equivalence, we require a few extra conditions.

41 Lemma With regard to the system (31), suppose that, corresponding to each set of input $v_i \in L_{2e}^{n_i}$, there exists a unique set $d_i \in L_{2e}^{n_i}$, $i=1,\ldots,k$ such that the equations (31) are satisfied. Suppose further that the operator mapping the input set $(v_1,\ldots,v_m)$ into $(d_1,\ldots,d_k)$ is casual. Then (32) holds if and only if the operator taking $(v_1,\ldots,v_m)$ into $(d_1,\ldots,d_k)$ has finite L_2-gain with zero bias.

Proof The "only if" part is obvious. To prove the "if" part, let $(v_1,\ldots,v_m)$ be a set of inputs in L_{2e}^n, and let $T < \infty$ be specified. Consider the system of equations

42a
$$d_{0i} = v_{iT} - \sum_{j=1}^{m} H_{ij}^{*} z_{0i}$$

42b
$$z_{0i} = G_i^{*} d_{0i} \qquad i=1,\ldots,m$$

By uniqueness and causality, we have $(d_{0i})_T = d_{iT}$, for $i=1,\ldots,k$. Also, since the operator taking $(v_1,\ldots,v_m)$ into $(d_1,\ldots,d_k)$ has finite L_2-gain with zero bias, there exists a finite constant ζ such that

43
$$\|d_{0i}\|_2 \le \zeta \sum_{j=1}^{m} \|v_{jT}\|_2 , \qquad i=1,\ldots,k$$

Finally, from (43) we get

44
$$\|d_i\|_{T2} = \|d_{0i}\|_{T2} \le \|d_{0i}\|_2 \le \zeta \sum_{j=1}^{m} \|v_j\|_{T2}, \quad i=1,\ldots,k$$

and the lemma is proved. □

9.2 SPECIAL CASES

In this section, we present some specific criteria for L_∞-stability, based on exponential weighting. These criteria are obtained by applying Theorem (9.1.29) in conjunction with the results of Chapters 6 and 7.

First, we consider a system described by

1a
$$e_i = u_i - H_i y_i - \sum_{j=1}^{m} B_{ij} y_j$$

1b
$$y_i = G_i e_i \qquad i=1,\ldots,m$$

where for all i we have $u_i, e_i, y_i \in L_{2e}^{n_i}$, and moreover

(i) G_i is a linear convolution operator with impulse response matrix $G_i(\cdot)$.

(ii) H_i is a memoryless nonlinearity in the sector $[c_i - (1-\delta_i)r_i,\ c_i+(1-\delta_i)r_i]$ for some real c_i and some positive δ_i, r_i.

(iii) For some $\alpha > 0$, the function $t \to G_i(t)\exp(\alpha t) \in L_2^{n_i \times n_i}$, and

(iv) $\sup_w E_{max}[F_i^\dagger(j\omega)\, F_i(j\omega)] \le r_i^{-2}$

where $E_{max}(\cdot)$ denotes the largest eigenvalue, $\dagger$ denotes the conjugate transpose, and

$F_i(j\omega) = [I + c_i\, \hat{G}_i(\alpha + j\omega)]^{-1}\, \hat{G}_i(\alpha+j\omega)$.

One can think of the system (1) as a collection of m single-loop feedback systems that are interacting through the operators B_{ij}, which may be nonlinear and time-varying.

Our objective is to derive some sufficient conditions for the L_∞-stability of the system (1), using Theorem (9.1.29) in conjunction with the results of Chapter 6. As a first step, observe that if $B_{ij} = 0\ \forall i,j$, then the system (1) is just a collection of m isolated subsystems, each of which can be shown to be L_∞-stable by virtue of (2) and Theorem (9.1.29). Hence it is intuitively clear that if the interactions B_{ij} are "sufficiently weak", then the system (1) is L_∞-stable. Theorem (17) below makes this idea precise. Before proceeding to the theorem, we give a preliminary

3 <u>Lemma</u> Suppose $f, q \in L_{2e}$ satisfy

4 $$q = f - H_i^* G_i^* q$$

Then

5 $$\|q\|_{T2} \le \eta_i \|f\|_{T2}$$

where

$$\eta_i = (|c_i| r_i^{-1} + 1)/\delta_i$$

<u>Proof</u> (4) is equivalent to

7 $$q = f - H_i^* q_1 \,, \quad q_1 = G_i^* q$$

Now define

8 $$q_2 = q + c_i q_i$$

Then in terms of q_2, (7) becomes

9 $$q_2 = f - (H_i^* - c_i I) q_1, \quad q_1 = G_i^* q_2 - c_i G_i^* q_1$$

or, equivalently,

10 $$q_2 = f - (H_i^* - c_i I)\, q_1, \quad q_1 = (I + c_i G_i^*)^{-1}\, G_i^* q_2$$

where the invertibility of $I + c_i G_i^*$ is guaranteed by (2) (note that G_i^* is a linear convolution operator with transfer function $\hat{g}(\alpha+s)$). Now from (10) we get

11 $$\|q_2\|_{T2} \le \|f\|_{T2} + \zeta_2[(H_i - c_i^* I)]\ \|q_1\|_{T2}$$

12 $$\|q_1\|_{T2} \le \zeta_2[(I + c_i G_i^*)^{-1}\ G_i^*]\ \|q_2\|_{T2}$$

Next, observe that G_i^* is a linear convolution operator whose transfer function matrix is $\hat{G}_i(\alpha+j\omega)$; hence (2) implies that $\zeta_2[(I+c_i C_i^*)^{-1}\ G_i^*] \le r_i^{-1}$ (see Lemma (3.1.83)).

Also, H_i^* belongs to the same sector as H_i, namely $[c_i - (1-\delta_i) r_i,\ c_i + (1+\delta_i) r_i]$, because of (9.1.11); hence $H_i^* - c_i I$ belongs to the sector $[-(1-\delta_i) r_i,\ (1+\delta_i) r_i]$, and $\zeta_2(H_i^* - c_i I) \le (1 + \delta_i) r_i$ (see Lemma (3.1.103)). Substituting these bounds into (11) and (12) and doing a little manipulation, we get

13 $$\|q_2\|_{T2} \le \|f\|_{T2} + (1-\delta_i)\ \|q_2\|_{T2}$$

14 $$\|q_2\|_{T2} \leq \delta_i^{-1} \; \|f\|_{T2}$$

15 $$\|q_2\|_{T2} \leq r_i^{-1}\delta_i^{-1} \; \|f\|_{T2}$$

where the last step follows by substituting (14) into (12). Finally, we have

16 $$\|q\|_{T2} \leq \|q_2\|_{T2} + |c_i| \; \|q_1\|_{T2}$$
$$\leq (\delta_i^{-1}+|c_i|\delta_i^{-1}r_i^{-1}) \; \|f\|_{T2}$$

which is the same as (5). □

Now we present the main stability criterion for the system (1).

17 <u>Theorem</u> Consider the system (1), and define the test matrix $P \in R^{m\times m}$ by

18 $$p_{ij} = \delta_{ij} - \eta_i \; \delta_2 \; (B_{ij}^* \; G_j^*)$$

where δ_{ij} denotes the Kronecker delta, η_i is given by (5), and

19 $$\zeta_2(B_{ij}^* G_j^*) = \sup_T \; \sup_{x_T\neq 0} \; \frac{\|B_{ij}^* G_j^* x\|_{T2}}{\|x\|_{T2}}$$

Under these conditions, the system (1) is L_∞-stable if the leading principal minors of P are all positive.

<u>Proof</u> In order to apply Theorem (9.1.29), we rewrite the system equations (1) as

20a $$e_i = u_i - \sum_{j=1}^{m} B_{ij} \; y_j - y_{i+m}$$

20b $$e_{i+m} = y_i$$

20c $$y_i = G_i \; e_i$$

$$\left.\right\} i=1,\ldots,m$$

20d $$y_{i+m} = H_i\, e_{i+m}$$

It is now routine to verify that the system (20) satisfies conditions (i)-(iii) of Theorem (9.1.29). Further, the "associated system" corresponding to (20), as given by (9.1.31), is

21a $$d_i = v_i - \sum_{j=1}^{m} B_{ij}^{*}\, z_j - z_{i+m}$$

21b $$d_{i+m} = z_i$$

21c $$z_i = G_i^{*}\, d_i$$

21d $$z_{i+m} = H_i^{*}\, d_{i+m}$$

$$\left.\right\} i=1,\ldots,m$$

It is easy to eliminate $z_1,\ldots,z_{2m}$ and $d_{m+1},\ldots,d_{2m}$ from (21); this gives

22 $$d_i = v_i - \sum_{j=1}^{m} B_{ij}^{*} G_j^{*} d_j - H_i^{*} G_i^{*} d_i, \quad i = 1,\ldots,m$$

Applying Lemma (3) to (22) gives

23 $$\begin{aligned}\|d_i\|_{T2} &\le \eta_i \|v_i - \sum_{j=1}^{m} B_{ij}^{*}\, G_j^{*}\, d_j\|_{T2} \\ &\le \eta_i \|v_i\|_{T2} + \sum_{j=1}^{m} \eta_i \gamma(B_{ij}^{*} G_j^{*})\, \|d_j\|_{T2}, \quad i=1,\ldots,m\end{aligned}$$

Hence (23) implies that

24 $$\sum_{j=1}^{m} P_{ij} \|d_j\|_{T2} \le \eta_i \|v_i\|_{T2}, \quad i=1,\ldots,m$$

Since P has all nonpositive off-diagonal elements, the condition that the leading principal minors of P are all positive implies, by Lemma (6.2.9), that P^{-1} exists and has all nonnegative elements. Hence from (24) we get

25 $$\|d_i\|_{T2} \le \sum_{j=1}^{m} (P^{-1})_{ij}\, \eta_j\, \|v_j\|_{T2}\,, \quad i=1,\ldots,m$$

Thus (9.1.32) is satisfied, whence the system (1) is L_∞-stable, by Theorem (9.1.29). □

Next, we present a typical passivity-type stability criterion for L_∞-stability. Clearly, by applying Theorem (9.1.29) in conjunction with each of the various results in Chapter 7, it is possible to generate a large number of such criteria. In the interests of brevity, we present only one such result, which corresponds to Theorem (7.3.28). We begin with some preliminary results.

26 Lemma Let $G: L_{2e} \to L_{2e}$ be a convolution operator with impulse response $g(\cdot)$, and suppose $t \to g(t)\exp(\alpha t) \in L_2$ for some $\alpha > 0$. If we denote $W_\alpha G W_\alpha^{-1}$ by G^*, then we have

27 $$\langle x, G^* x\rangle_T \ge \varepsilon\, \|x\|_{T2}^2, \; \forall T \ge 0, \; \forall x \in L_{2e}$$

where

28 $$\varepsilon = \min_{\omega \in R} \operatorname{Re} \hat{g}(\alpha + j\omega)$$

Proof This is a special case of Lemma (3.2.19).

29 Lemma Let $\phi: R \to R$ belong to the sector [0,k], i.e.

30 $$0 \le \sigma\phi(\sigma) \le k\sigma^2, \quad \forall \sigma \in R$$

and let $G: L_{2e} \to L_{2e}$ be defined by

31 $$(Gx)(t) = \phi(x(t))$$

Let $\alpha > 0$ and let $G^* = W_\alpha G W_\alpha^{-1}$. Then we have

32 $$\langle x, G^* x\rangle_T \ge (1/k)\, \|Gx\|_{T2}^2\,, \quad \forall T \ge 0, \quad \forall x \in L_{2e}$$

<u>Proof</u> We have

33 $$\langle x, G^* x\rangle_T = \int_0^T x(t)\exp(\alpha t)\,\phi[x(t)\exp(-\alpha t)]\,dt$$

$$= \int_0^T \exp(2\alpha t)\cdot x(t)\exp(-\alpha t)\phi[x(t)\exp(-\alpha t)]\,dt$$

$$\geq (1/k)\int_0^T \exp(2\alpha t)(\phi[x(t)\exp(-\alpha t)])^2 dt$$

$$= (1/k)\int_0^T [(G^* x)(t)]^2 dt$$

$$= (1/k)\,\|G^* x\|_{T2}$$

so that (32) is proved. □

34 <u>Theorem</u> Consider the system

35a $$e_i = u_i - \sum_{j=1}^{m} h_{ij} y_j$$

35b $$y_i = G_i e_i$$

$\Big\}$ $i=1,\ldots,m$

where u_i, e_i, $y_i \in L_{2e}$ for all i, h_{ij} are real constants, and G_i: $L_{2e} \to L_{2e}$. Suppose that for some integer $k \leq m$, G_i is a convolution operator with impulse response $g_i(\cdot)$ for $i=1,\ldots,k$; in addition, for $i = k+1,\ldots,m$, G_i is a memoryless time-invariant nonlinearity in the sector $[0,k_i]$. Suppose $h_{ij} = 0$ f for $k+1 \leq i,j \leq m$. Finally, suppose that for some $\alpha > 0$, the functions $t \to g_i(t)\exp(\alpha t) \in L_2$ for $i=1,\ldots,k$. Now define

36 $$\varepsilon_i = \min_{\omega \in R} \hat{g}_i(\alpha + j\omega), \quad i=1,\ldots,k$$

37 $$\delta_i = 1/k_i, \quad i = k+1,\ldots,m$$

38 $$E_a = \text{Diag}\,\{\varepsilon_1,\ldots,\varepsilon_k\}$$

39 $$D_b = \text{Diag}\,\{\delta_{k+1},\ldots,\delta_m\}$$

40
$$H = (h_{ij}) = \begin{bmatrix} H_{aa} & H_{ab} \\ H_{ba} & 0 \end{bmatrix} \begin{matrix} k \\ m-k \end{matrix}$$
$$\qquad\qquad\qquad\quad k \qquad m-k$$

41
$$M = \begin{bmatrix} H'_{aa} E_a H_{aa} & H'_{aa} E_a H_{ab} \\ H'_{ab} E_a H'_{aa} & H'_{ab} E_a H_{ab} + D_b \end{bmatrix}$$

Under these conditions, the system (35) is L_∞-stable if

(i) M is positive semidefinite

(ii) we have

42
$$\text{rank } M = \text{rank } \left[M \,\middle|\, \begin{matrix} H'_{aa} \\ H'_{ab} \end{matrix} \right]$$

(iii) $H'_{ab} E_a H_{ab} + D_b$ is positive definite

(iv) H is positive

Proof Apply Theorem (9.1.29) with Theorem (7.3.28). □

9.3 GENERAL INSTABILITY RESULT

In this section, we derive a general instability result, that permits one to deduce the L_∞-instability of a given system by establishing the L_2-instability of a associated system. In this sense, this result is an instability counterpart of Theorem (9.1.29).

We begin by establishing a preliminary result.

1 Lemma Suppose $G(\cdot)$ is an n×n-dimensional Laplace transformable distribution with support in $[0,\infty)$, and define the corresponding operator G: $L^n_{2e} \to L^n_{2e}$ by

2 $$(Gx)(t) = \int_0^t G(t-\tau)\, x(\tau)\, d\tau$$

Suppose G belongs to class U (see Definition (3.3.10)), and that $\hat{G}(\cdot)$ has a right-coprime factorization $(\hat{N}(\cdot), \hat{D}(\cdot))$ in $\hat{A}^{n\times n}$ (see Definition (3.3.15)) where $D(\cdot)$ is of the form $D(t) = D_o\delta(t) + D_a(t)$, $D_a(\cdot) \in L_1^{n\times n}$. Under these conditions,

3 $$\text{(i)} \quad \sup_{\omega} \|\hat{G}(j\omega)\| < \infty$$

(ii) The only singularities of $\hat{G}(\cdot)$ in C_+ are isolated poles in C_+^o of finite order.* Moreover, there is a $\sigma_o > 0$ such that $G(\cdot)$ has no singularities in the strip $0 \le \text{Re } s \le \sigma_o$.

Proof (i) follows readily from Lemma (3.3.48), and the fact that $\zeta_{c2}(G) < \infty$. To prove (ii), let $(\hat{N},\hat{D})$ be the r.c.f. of $\hat{G}$. Then

4 $$\inf_{\omega} |\det \hat{D}(j\omega)| > 0$$

Otherwise, there exists a sequence $\{\omega_i\}$ such that Det $\hat{D}(j\omega_i) \to 0$, whence $\hat{G}(j\omega_i)$ becomes unbounded, in violation of (3) (see Lemma (3.3.28)). Now, consider the function

5 $$f(\sigma,M) = \sup_{|\omega| \le M} |\det \hat{D}(\sigma+j\omega)|$$

By (4), $f(0,M) > 0$. Since $f(\cdot,M)$ is continuous, for every M there is a $\sigma_o(M) > 0$ such that $f(\sigma,M) > 0$ whenever $0 \le \sigma \le \sigma_o(M)$. Also, by the Riemann-Lebesgue Lemma, $\hat{D}(\sigma+j\omega) \to D_o$ whenever $\sigma \ge 0$ and $|\omega| \to \infty$; moreover, $\det D_o \ne 0$ by (4). Combining these facts, we see that there exists a $\sigma_o > 0$ such that

6 $$\inf_{\omega} \inf_{0\le\sigma\le\sigma_o} |\det \hat{D}(\sigma+j\omega)| > 0$$

---

* Recall that $C_+ = \{s: \text{Re } s \ge 0\}$ and $C_+^o = \{s; \text{Re } s > 0\}$

This shows that $\hat{G}(\cdot)$ has no singularities in the strip $0 \le \text{Re } s \le \sigma_o$, i.e. that all singularities of $\hat{G}$ must lie in the half-plane $\text{Re } s > \sigma_o$. Since $\hat{D}$ is analytic in C_+^o, all of these singularities must be isolated poles of finite order. □

Lemma (1) shows that if G belongs to class U and $\hat{G}$ has an r.c.f. in $A^{n\times n}$, then it is reasonable to assume that $\hat{G}$ has a pole at some $s_o \in C_+^o$. Further, if $\hat{G}$ has a pole at $s_o \in C_+^o$, then one can explicitly calculate some elements of $M^{\perp}(G)$, using Theorem (3.3.45). This is the key point in the proof of Theorem (13) below, which is the main result of this section.

Consider now a large-scale system described by

7a, 7b

$$\left.\begin{aligned} e_i &= u_i - \sum_{j=1}^{m} H_{ij} y_j \\ y_i &= G_i\, e_i \end{aligned}\right\} \quad i=1,\dots,m$$

where $e_i, u_i, y_i \in L_{2e}^{n_i}$ for some integer n_i, G_i: $L_{2e}^{n_i} \to L_{2e}^{n_i}$, and H_{ij}: $L_{2e}^{n_j} \to L_{2e}^{n_i}$, $\forall i,j$. To set the stage for a general L_∞-instability result for the system (7), we begin with a discussion of L_2-instability. There are several diverse criteria for L_2-instability given in Sections 8.1, 8.2, and 8.3.2, but they all share two common features: (i) they assume that some of the operators, say $G_1,\dots,G_k$, belong to class U; (ii) they state that, if appropriate instability criteria are satisfied, then L_2-instability occurs* whenever the input u_i in (7) are selected to be of the form

8

$$u_i = \sum_{j=1}^{k} F_{ij} v_j, \qquad v_j \in M^{\perp}(G_j)/\{0\}$$

where the F_{ij} are constant matrices that depend on the particular criteria used.

We now introduce the exponential weighting used to

---

* i.e., either $e_i \notin L_2^{n_i}$ or $y_i \notin L_2^{n_i}$ for some i.

establish the L_∞-instability of the system (7) by studying the L_2-instability of an associated system. Given $\lambda > 0$, recall from Definition (9.1.1) that the operator W_λ: $L_{2e}^n \to L_{2e}^n$ is defined by

9 $$(W_\lambda x)(t) = x(t) \exp(\lambda t)$$

clearly $W_\lambda^{-1} = W_{-\lambda}$ also maps L_{2e}^n into itself. Now, given the system (7), we can rewrite (7) as

10a $$W_{-\lambda} e_i = W_{-\lambda} u_i - \sum_{j=1}^{m} (W_{-\lambda} H_{ij} W_\lambda) \cdot W_{-\lambda} y_j \qquad i=1,\ldots,m$$

10b $$W_{-\lambda} y_i = (W_{-\lambda} G_i W_\lambda) W_{-\lambda} e_i$$

Note that, if $\lambda > 0$, then the above technique of exponential weighting is the <u>opposite</u> of that proposed in Section 9.1. For the sake of brevity, let us define

11 $$G_i^\# = W_{-\lambda} G_i W_\lambda \quad , \quad H_{ij}^\# = W_{-\lambda} H_{ij} W_\lambda$$

Then the "associated system" corresponding to (7) is

12a $$d_i = r_i - \sum_{j=1}^{m} H_{ij}^\# z_j$$

12b $$z_i = G_i^\# d_i$$

$$\left.\right\} \; i=1,\ldots,m$$

where r_i is the new "input", d_i is the new "error", and z_i is the new "output". Moreover, $r_i, d_i, z_i \in L_{2e}^{n_i}$, and $G_i^\#: L_{2e}^{n_i} \to L_{2e}^{n_i}$ and $H_{ij}^\#: L_{2e}^{n_j} \to L_{2e}^{n_i}$, $\forall i,j$. Theorem (13) below relates the L_∞-instability of the <u>original</u> system (7) to that of the <u>associated</u> system (12).

13 <u>Theorem</u> Consider the system (7), and suppose the following conditions are satisfied:

(c1) For some integer $k \leq m$, the operators G_i, $i=1,\ldots,k$ satisfy the hypotheses that (i) G_i is a convolution operator belonging to class U, (ii) $\hat{G}_i$ has an r.c.f. in $\hat{A}^{n_i \times n_i}$, and $\hat{G}_i$ has a pole at some $s_i \in C_+^o$, for i=1,...,k.

(c2) Select $\lambda > 0$ such that

14 $$\sup_{\omega} \|\hat{G}_i(\lambda+j\omega)\| < \infty, \quad i=1,\ldots,k$$

and such that G_i has a pole $s_{io} = \sigma_{io} + j\omega_{io}$ with $\sigma_{io} < 2\lambda$ (see Remark (i) below). For this choice of λ, consider the associated system (12). Suppose the system (12) is L_2-unstable, and further, there exist matrices F_{ij}, i=1,...,m, j=1,...,k such that L_2-instability results i.e. $d_i \notin L_2^{n_i}$ or $z_i \notin L_2^{n_i}$ for some i whenever

15 $$r_i = \sum_{j=1}^{k} F_{ij} v_j, \quad v_j \in M^{\perp}(G_j^{\#})/\{0\}, \; j=1,\ldots,k, \; i=1,\ldots,m$$

Under these conditions, the original system (7) is L_∞-unstable.

<u>Remarks</u> (i) It is always possible to select λ so as to satisfy (14). In view of (c1) and Lemma (1), there exists a $\sigma_o > 0$ such that none of the functions $\hat{G}_i(\cdot)$ has a singularity in the strip $0 \le \text{Re } s \le \sigma_o$; one can choose any λ in the interval $[0, \sigma_o/2]$. However, other choices for λ may also be possible. Note that $\hat{G}_i(\cdot)$ is only required to have <u>one</u> pole with real part greater than 2λ; there is no requirement that <u>all</u> poles of $\hat{G}_i(\cdot)$ have real parts greater than 2λ.

(ii) As mentioned earlier, the requirement that L_2-instability of (12) results for inputs r_i of the form (15) is not at all restrictive, since all of the instability criteria of Sections 8.1, 8.2, and 8.3.2 meet this requirement.

<u>Proof</u> First of all, note that if G_i is a convolution operator with kernel $\hat{G}_i(\cdot)$, then $G_i^{\#}$ is also a convolution operator with $G_i(\cdot)\exp(-\lambda(\cdot))$, so that $\hat{G}_i^{\#}(s) = \hat{G}_i(s+\lambda)$. Also, $G_i^{\#}$ belongs to class U, for i=1,...,k. To see this, note that $G_i^{\#}$ is linear, and satisfies (3.3.11) (because $W_{-\lambda}$, G_i, W_λ

individually satisfy (3.3.11)); finally, (14) guarantees that $\zeta_c(G_i^{\#}) < \infty$.

It is easy to see that if $(\hat{N}_i, \hat{D}_i)$ is an r.c.f. of $\hat{G}_i$, then $(\hat{N}_i(s+\lambda), \hat{D}_i(s+\lambda)$ is an r.c.f. of $\hat{G}_i(s+\lambda)$. Similarly, if $\hat{G}_i$ has a pole at s_{io}, then $\hat{G}_i^{\#}$ has a pole at $s_{io}-\lambda$. By Theorem (3.3.45), $M^{\perp}(G_i^{\#})$ contains a nonzero element of the form $v_i \exp[(-\sigma_{io}+\lambda)t] \cos(\omega_{io}t)$. Now, by condition (c2), if we select

16 $$r_i(t) = \sum_{j=1}^{k} F_{ij} v_j \exp[(-\sigma_{jo}+\lambda)t]\cos(\omega_{jo}t), \; i=1,\ldots,m$$

in (12), then either $z_i \notin L_2^{n_i}$ or $d_i \notin L_2^{n_i}$ for some i. Comparing (10) and (12), we see that if we select

17 $$W_{-\lambda}u_i(t) = \sum_{j=1}^{k} F_{ij}v_j \exp[(-\sigma_{jo}+\lambda)t]\cos(\omega_{jo}t), \; i=1,\ldots,m$$

i.e., if we select

18 $$u_i(t) = \sum_{j=1}^{k} F_{ij}v_j \exp[(-\sigma_{jo}+2\lambda)t] \cos(\omega_{jo}t), \; i=1,\ldots,m$$

in (7), then either $W_{-\lambda}y_i \notin L_2^{n_i}$ or $W_{-\lambda}e_i \notin L_2^{n_i}$ for some i. In particular, this means that, with the choice of inputs $u_i(\cdot)$ of (18), either $y_i \notin L_\infty^{n_i}$ or $e_i \notin L_\infty^{n_i}$ for some i (note that, if $f(\cdot) \in L_\infty^{n_i}$, then $W_{-\lambda} f(\cdot) \in L_2^{n_i}$). Since the input u_i defined by (18) belongs to $L_\infty^{n_i}$, we see that the system (7) is L_∞-unstable.□

It is obvious that, using Theorem (9.3.13), one can obtain L_∞-instability criteria corresponding to almost all of the L_2-instability criteria of sections 8.1, 8.2, and 8.3.2. It is unnecessary to list of all of these. However, we state a "small gain" type result for the purposes of illustration.

19 <u>Theorem</u> Consider a single-loop feedback system described by

20 $$e_1 = u_1-y_2, \; e_2 = u_2+y_1, \; y_1 = G_1e_1, \; y_2 = G_2e_2$$

where G_1 is a convolution operator of class U, and has an r.c.f. in $A^{n\times n}$; G_2 is a memoryless operator of the form

21 $$(G_2 e_2)(t) = \phi(t, e_2(t))$$

and ϕ belongs to the sector $[\alpha, \beta]$, i.e.,

22 $$[\phi(t,v) - \alpha v]' [\phi(t,v) - \beta v] \le 0, \quad \forall t \ge 0, \ \forall v \in R^n$$

under these conditions, the system (20) is L_∞-unstable if one can find a $\lambda > 0$ such that the following conditions are satisfied:

23 $$\text{(i)} \quad \sup_\omega E_{max} [\hat{H}^\dagger(\lambda+j\omega) \hat{H}(\lambda+j\omega)] < \delta^{-1}$$

24 (ii) $\hat{H}(\cdot)$ has at least one pole with real part greater than 2λ, where

25 $$\hat{H}(s) = \hat{G}(s) [I + \gamma \hat{G}(s)]^{-1}$$

26 $$\gamma = [\beta+\alpha]/2, \quad \delta = [\beta-\alpha]/2$$

and $E_{max}(\cdot)$ denotes the largest eigenvalue of a matrix, $\dagger$ denotes the conjugate transpose.

The proof is simple and is therefore omitted.

Comparing Theorem (19) with Theorem (9.2.17), we see the former involves $G_i(\lambda+j\omega)$ for some $\lambda > 0$, while the latter involves $G_i(-\alpha+j\omega)$ for some $\alpha > 0$.

NOTES AND REFERENCES

The exponential weighting approach to study L_∞-stability is due to Sandberg [San. 2] and Zames [Zam. 2]. Related results can be found in [Ber. 1]. This approach was extended to large-scale systems by Lasley and Michel [Las. 2]. The general stability theorem given here is taken from [Vid. 6], while specific results are from [Las. 2]. The approach to L_∞-instability is contained in [Vid. 8].

REFERENCES

Ara.1 Araki, M. , "Application of M-Matrices to the Stability Problems of Composite Dynamical Systems", J. Math. Anal. Appl., Vol. 52, pp. 309-321, Nov. 1975.

Ara.2 Araki, M., "Input-Output Stability of Composite Feedback Systems", IEEE Trans. Auto. Control, Vol. AC-21, pp. 254-259, April 1976.

Ara.3 Araki, M., "Stability of Large-Scale Nonlinear Systems-Quadratic Order Theory of Composite System Method Using M-Matrices", IEEE Trans. Auto. Control, Vol. AC-23, pp. 129-142, April 1978.

Baa.1 Baase, S., Computer Algorithms: Introduction to Design and Analysis, Addison-Wesley, Reading, MA, 1978.

Ber.1 Bergen, A.R., Iwens, R. and Rault, A.J., "On Input-Output Stability of Nonlinear Feedback Systems", IEEE Trans. Auto. Control, Vol. AC-11, pp. 742-745, Oct. 1966.

Cal.1 Callier, F.M., Chan, W.S., and Desoer, C.A., "Input-Output Stability Theory of Interconnected Systems Using Decomposition Techniques", IEEE Trans. Circ. and Sys., Vol. CAS-23, pp. 714-729, December 1976.

Cal.2 Callier, F.M., Chan, W.S., and Desoer, C.A., "Input-Output Stability of Interconnected Systems Using Decompositions: An Improved Formulation", IEEE Trans. Auto. Control, Vol. AC-23, pp. 150-163, April 1978.

Cho.1 Cho, Y.S., and Narendra, K.S., "Stability of Nonlinear Time-Varying Feedback Systems", Proc. 5th. Allerton Conf., pp. 249-258, Oct. 1967.

Cho.2 Cho, Y.S., and Narendra, K.S., "Stability of Nonlinear Time-Varying Feedback Systems", Automatica, Vol. 4, pp. 309-322, Nov. 1968.

Coo.1 Cook, P.A., "On the Stability of Interconnected Systems", Int. J. Control, Vol. 20, pp. 407-416, Sept. 1974.

Des.1 Desoer, C.A., and Wu, M.-Y., "Stability of Linear Time-Invariant Systems", IEEE Trans. Circ. Thy., Vol. CT-15, pp. 245-250, Sept. 1968.

Des.2 Desoer, C.A. and Vidyasagar, M., Feedback Systems: Input-Output Properties, Academic Press, New York, 1975.

Dun.1 Dunford, N., and Schwartz, J.T., Linear Operators, Part I, Interscience, New York, 1959.

Fie.1 Fiedler,M., and Ptak,V., "On Matrices with Nonpositive Off-diagonal Elements and Positive Principal Minors", Czech. Math. J., Vol. 12, pp. 382-400, 1962.

Fre.1 Freeman, H., Discrete-Time Systems, Wiley, New York, 1965.

Hil.1 Hill,D.J., and Moylan,P.J., "The Stability of Nonlinear Dissipative Systems", IEEE Trans. Auto. Control, Vol. AC-21, pp. 708-711, Oct. 1976.

Hil.2 Hill,D.J., and Moylan,P.J., "Stability Results for Nonlinear Feedback Systems", Automatica, Vol. 13, pp. 377-382, July 1977.

Hil.3 Hill,D.J., and Moylan,P.J., "A General Result on the Instability of Feedback Systems", Proc. Int'l. Symp. on Circ. and Sys., pp. 652-657, April 1978.

Las.1 Lasley,E.L., and Michal,A.N., "Input-Output Stability of Interconnected Systems", IEEE Trans. Auto. Control, Vol. AC-21, pp. 84-89, February 1976.

Las.2 Lasley,E.L., and Michel,A.N., "L_∞- and ℓ_∞-stability of Interconnected Systems", IEEE Trans. Circ. and Sys., Vol. CAS-23, pp. 261-270, May 1976.

Mic.1 Michel,A.N., and Miller,R.K., Qualitative Analysis of Large-Scale Dynamical Systems, Academic Press, New York, 1977.

Moy.1 Moylan,P.J., "Matrices with Positive Principal Minors", Linear Algebra and Its Appl., Vol. 17, pp. 53-58, 1977.

Moy.2 Moylan,P.J., and Hill,D.J., "Stability Criteria for Large-Scale Systems", IEEE Trans. Auto. Control, Vol. AC-23, pp. 143-150, April 1978.

Moy.3 Moylan,P.J., and Hill,D.J., "Tests for Stability and Instability of Interconnected Systems", IEEE Trans. Auto Control, Vol. AC-24, pp. 574-575, Aug. 1979.

Moy.4 Moylan,P.J., Vannelli,A., and Vidyasagar,M., "On the Stability and Well-Posedness of Interconnected Nonlinear Dynamical Systems", IEEE Trans. Circ. and Sys., Vol. CAS-27, Nov. 1980.

Por.1 Porter,D.W., and Michel,A.N., "Input-Output Stability of Time-Varying Nonlinear Multiloop Feedback Systems", IEEE Trans. Auto. Control, Vol. AC-19, pp. 422-427, August 1974.

Sae.1 Saeks, R., Resolution Space, Operators, and Systems, Springer-Verlag, Berlin, 1973.

San.1 Sandberg, I.W., "On the L_2-Boundedness of Solutions of Nonlinear Functional Equations", Bell Sys. Tech. J., Vol. 43, pp. 1581-1599, July 1964.

San.2 Sandberg, I.W., "Some Results on the Theory of Physical Systems Governed by Nonlinear Functional Equations", Bell Sys. Tech. J., Vol. 44, pp. 871-898, May-June 1965.

San.3 Sandberg, I.W., "On the Stability of Linear Systems Containing a Time-Varying Element with a Restricted Rate of Variation", IEEE Int'l. Conv. Rec., Vol. 14, Part 7, pp. 173-183, March 1966.

San.4 Sandberg, I.W., "On the Stability of Interconnected Systems", Proc. Int'l. Symp. on Circ. and Sys., pp. 228-231, April 1978.

Sil.1 Siljak, D.D., Large-Scale Dynamic Systems, Elsevier-North-Holland, 1978.

Sun.1 Sundareshan, M.K., and Vidyasagar, M., "L_2-Stability of Large-Scale Dynamical Systems-Criteria via Positive Operator Theory", IEEE Trans. Auto. Control, Vol. AC-22, pp. 396-399, June 1977.

Tak.1 Takeda, S., and Bergen, A.R., "Instability of Feedback Systems by Orthogonal Decomposition of L_2", IEEE Trans. Auto. Control, Vol. AC-18, pp. 631-636, June 1973.

Tar.1 Tarjan, R., "Depth-First Search and Linear Graph Algorithms", SIAM J. Comp., Vol. 1, pp. 146-160, June 1972.

Vid.1 Vidyasagar, M., "Coprime Factorizations and Stability of Multivariable Distributed Feedback Systems", SIAM J. Control, Vol. 13, pp. 1144-1155, November 1975.

Vid.2 Vidyasagar, M., "L_2-Instability Criteria for Interconnected Systems", SIAM J. Contr. and Opt., Vol. 15, pp. 312-328, February 1977.

Vid.3 Vidyasagar, M., "L_2-Stability of Interconnected Systems Using a Reformulation of the Passivity Theorem", IEEE Trans. on Circ. and Sys., Vol. CAS-24, pp. 637-645, November 1977.

Vid.4 Vidyasagar, M., Nonlinear Systems Analysis, Prentice-Hall, Englewood Cliffs, N.J., 1978.

Vid.5 Vidyasagar,M., "On the Use of Right Coprime Factorizations in Distributed Feedback Systems Containing Unstable Subsystems", IEEE Trans. on Circuits and Sys., Vol. CAS-25, pp. 916-921, November 1978.

Vid.6 Vidyasagar,M., "L_∞-Stability Criteria for Interconnected Systems using Exponential Weighting", IEEE Trans. Circ. and Sys., Vol. CAS-25, pp. 946-947, Nov. 1978.

Vid.7 Vidyasagar,M., "Existence of Right-Coprime Factorizations and L_∞-Instability of Large-Scale Interconnected Systems", Proc. Symp. Math. Thy. of Netw. and Sys., Delft, The Netherlands, pp. 303-309, July 1979.

Vid.8 Vidyasagar,M., "New Passivity Type Criteria for Large-Scale Interconnected Systems", IEEE Trans. Auto. Control, Vol. AC-24, pp. 575-579, August 1979.

Vid.9 Vidyasagar,M., "On the Well-Posedness of Large-Scale Interconnected Systems", IEEE Trans. Auto. Control, Vol. AC-25, pp. 413-420, June 1980.

Vid.10 Vidyasagar,M., "New L_2-Instability Criteria for Large-Scale Interconnected Systems, "IEEE Trans. Circ. and Sys., Vol. CAS-27, Oct. 1980.

Wil.1 Willems,J.C., "Some Results on the L^p-Stability of Time-Varying Linear Systems", IEEE Trans. Auto. Control, Vol. AC-14, pp. 660-665, December 1969.

Wil.2 Willems,J.C., The Analysis of Feedback Systems, MIT Press, Cambridge, MA, 1971.

Wil.3 Willems,J.C., "The Construction of Lyapunov Functions for Input-Output Stable Systems", SIAM J. Control, Vol. 9, pp. 105-134, February 1971.

Wil.4 Willems,J.C., "Dissipative Dynamical Systems, Part I: General Theory; Part II: Linear Systems with Quadratic Supply Rates", Arch. Rat. Mech. Anal., Vol. 45, pp. 321-393, 1972.

Wil.5 Willems,J.C., "Mechanisms for the Stability and Instability in Feedback Systems", Proc. IEEE, Vol. 64, pp. 24-35, January 1976

Wu.1 Wu,M.-Y. and Desoer,C.A., "L^p-Stability ($1 \le p \le \infty$) of Nonlinear Time-Varying Feedback Systems", SIAM J. Control, Vol. 7, pp. 356-364, May 1969.

Zam.1 Zames,G., "On the Stability of Nonlinear Time-Varying Feedback Systems", Proc. NEC, Vol. 20, pp. 725-730, Nov. 1964.

Zam.2 Zames,G., "Nonlinear Time-Varying Feedback Systems-Conditions for L_∞-Boundedness Derived Using Conic Operators on Exponentially Weighted Spaces", Proc. 3rd. Allerton Conf., pp. 460-471, Oct. 1965.

Zam.3 Zames,G., "On the Input-Output Stability of Nonlinear Time-Varying Feedback Systems", Parts I and II, IEEE Trans. Auto. Control, Vol. AC-11, pp. 218-238, April 1966 and pp. 465-476, July 1966.

Zam.4 Zames,G. and Falb,P.L., "Stability Conditions for Systems with Monotone and Slope-Restricted Nonlinearities", SIAM J. Control, Vol. 6, pp. 89-108, Feb. 1968.

INDEX

W

Lecture Notes in Control and Information Sciences

Edited by A. V. Balakrishnan and M. Thoma

GPSR Compliance

The European Union's (EU) General Product Safety Regulation (GPSR) is a set of rules that requires consumer products to be safe and our obligations to ensure this.

If you have any concerns about our products, you can contact us on ProductSafety@springernature.com

In case Publisher is established outside the EU, the EU authorized representative is:

Springer Nature Customer Service Center GmbH
Europaplatz 3
69115 Heidelberg, Germany

Batch number: 10704763

Printed by Printforce, the Netherlands